KT-147-790

RELIABILITY AND SIX SIGMA

RELIABILITY AND SIX SIGMA

U Dinesh Kumar

Indian Institute of Management Calcutta, Kolkata, India

John Crocker

Data Systems and Solutions, Bristol, UK

T Chitra

Ex-scientist, Defence Research and Development Organization, Bangalore, India

Haritha Saranga

Indian School of Business, Hyderabad, India

U Dinesh Kumar
Indian Institute of Management
Kolkata, India

John Crocker
Data Systems and Solutions
Bristol, UK

T Chitra
Defence Research & Development Org.
Bangalore, India

Haritha Saranga
Indian School of Business
Hyderabad, India

Library of Congress Control Number: 2005935848

ISBN-10: 0-387-30255-7 (HB) ISBN-10: 0-387-30256-5 (e-book)
ISBN-13: 978-0387-30255-3 (HB) ISBN-13: 978-0387-30256-0 (e-book)

Printed on acid-free paper.

Printed in the United States of America.

9 8 7 6 5 4 3 2 1 SPIN 11054146

springeronline.com

Dedication

This book is dedicated to:

My wonderful partner Haritha and gorgeous daughter Prathusha

- U Dinesh Kumar

My wife Frances and daughters Elizabeth and Louise

- John Crocker

My parents Sarada & Thyagarajan
and mentors Dr S Krishnan and Dr N K Srinivasan

- T Chitra

My amazing partner Dinesh and adorable daughter Prathusha

- Haritha Saranga

Contents

CHAPTER 3: RELIABILITY AND SIX SIGMA MEASURES

CHAPTER 4: SYSTEM RELIABILITY

CHAPTER 5: DESIGN FOR RELIABILITY AND SIX SIGMA

CHAPTER 6: IN-SERVICE RELIABILITY

CHAPTER 7: RELIABILITY AND SIX SIGMA ESTIMATION

CHAPTER 10: RELIABILITY AND SIX SIGMA MANAGEMENT

Preface

If you have knowledge let others light their candles with it

Winston Churchill

Six Sigma is at the top of the agenda for many companies which try to reduce cost and improve productivity. Companies such as General Electric, Motorola, 3M implement thousands of Six Sigma projects every year. This is probably the only concept which has attracted the attention of almost all companies across the world irrespective of their business. Six Sigma was introduced by Motorola with an objective to reduce the defects and product failure rate. Within a short span of its existence Six Sigma has become a major strategy to run organizations. The concept which was introduced to solve the manufacturing problems has evolved into a strategy to run organizations and has become a basis for many business models. It is natural to ask why Six Sigma is so popular and whether it really delivers or is it just another fad.

Six Sigma is definitely different from other tools, but most components of Six Sigma are well known and were in use long before Six Sigma became popular. Six Sigma has created very few new tools to solve problems. The strength of Six Sigma is the way in which it structures the problem and the solution methodology to solve the problem. Its structured methodology and rigour are the main differentiators between Six Sigma and other process improvement techniques. However, Six Sigma does not guarantee that it will solve all business related problems. It is not a panacea; there is no 'one stop' solution to problems.

At the operational level Six Sigma is used to reduce the defects, however, reducing the defects alone cannot guarantee success of a product. Thanks to Google, today customers are probably more informed than ever before. The worst thing for any business is to deal with customers who know too much about the product. It has become hard to sell any consumer durable product unless it is the cheapest and the best in all system characteristics. Among the system characteristics, reliability is a top priority to almost all customers. If the product is not reliable, the company should gain knowledge in the scrap metal business, as it is likely to come in handy for its survival. This book is an attempt to combine Six Sigma and Reliability concepts and understand how these two concepts are interconnected.

The main objective of this book is to provide an integrated approach to Reliability and Six Sigma. For the first time we have introduced several mathematical models that relate Six Sigma and Reliability measures. Several case studies are used throughout the book to illustrate the application of the models discussed.

The additional objectives of the book are:

1. Introduce the concept of reliability and Six Sigma and their role in system effectiveness.
2. Introduce basic probability and statistical techniques that are essential for analyzing reliability and Six Sigma problems.
3. Introduce Six Sigma and reliability measures: how to predict them; how to interpret them and how reliability and Six Sigma measures are related.
4. Discuss tools and techniques for evaluation of system reliability and the relationship between system reliability and Sigma level.
5. Discuss design-for-reliability and Six Sigma and provide tools that are essential to implement Reliability and Six Sigma projects.
6. Discuss issues related to in-service reliability and how to manage in-service reliability.
7. Introduce techniques for estimation of reliability and Six Sigma from in-service and test data.
8. Introduce software reliability models and Six Sigma quality in software.
9. Discuss availability and Sigma level.
10. Discuss management of reliability and Six Sigma, DMAIC (Define-Measure-Analyze-Improve and Control) methodology.

In the first chapter we introduce the concept of Six Sigma and its contribution to organizational effectiveness. The relationship between reliability and Six Sigma along with their importance to organizational effectiveness is discussed in this chapter. In Chapter 2, we introduce the concept of probability, random variables and probability distributions. In particular, we will look at ways of describing time-to-failure random variable using distributions such as exponential, normal, Weibull and lognormal and how these can be used to provide better reliability models. The normal distribution is discussed in detail to provide the foundation for Six Sigma methodology.

Chapter 3 introduces Six Sigma and Reliability measures. Six Sigma measures such as Yield, DPMO (defects per million opportunities) and Sigma Level are discussed and mathematical models are derived which can be used to evaluate these measures. Several useful reliability measures including the failure function, reliability function, hazard function, mean time between failure (MTBF) and maintenance free operating period (MFOP) are also discussed. Mathematical models are derived to link Reliability and Six Sigma measures using Sigma Level Quality.

Chapter 4 deals with mathematical tools required for predicting reliability of series, parallel, series-parallel, complex and network systems. Many real life examples are used to illustrate the application of the methods discussed in the chapter.

Reliability is a design characteristic and any reliability project should start from the conceptual design stage of the product development. Similarly, Six Sigma can either be used as an operational strategy where the objective is to reduce defect and production cost or as a corporate strategy where the product development from the initial stage follows a Six Sigma program (called Design for Six Sigma, DFSS). Design for reliability and Six Sigma require advanced mathematical and management tools. Chapter 5 is dedicated to this aspect of reliability and Six Sigma. Managing in-service reliability is a nightmare for any original equipment manufacturer (OEM). Managing in-service reliability requires special tools and techniques which are discussed in Chapter 6.

Chapter 7 is dedicated to statistical tools such as regression and maximum likelihood estimator for determining the best fit distribution for the time-to-failure random variables and estimating their parameters from test as well as in-service data.

Models for software reliability prediction and estimation are discussed in Chapter 8. Software is a little bit like this book, we know it contains errors that will have originated from us, its authors. What we do not know is where they are or how many. With your help, we would hope to remove some of these errors with each reprint or new edition. But, alas, we may also introduce new ones. This makes the task of prediction particularly difficult.

Availability is a measure of the overall impact of reliability, maintainability and supportability of the product and can be classified as a true measure of design effectiveness. Different availability-related measures are discussed in Chapter 9 along with their relationships to Six Sigma.

The success of any process improvement concept will depend on how well it is managed and implemented. Management of reliability and Six Sigma projects is discussed in chapter 10. Most of this chapter is dedicated to the DMAIC cycle. The chapter also discusses issued like reliability demonstration, reliability growth and DFSS.

Most of the material presented in this book have been tested from time to time with different types of students, from technicians to top level management from various multi-national companies. The book is intended for Under-graduate and Post-graduate students from all engineering and management disciplines. Above all, the book is designed to be a useful reference for Reliability Engineers and Six Sigma practitioners including Six Sigma green and black belts from all types of industry and those people encumbered with process improvement and new product development.

As Alan Mulally, in the capacity of the General Manager of Boeing 777 division used to say frequently: *we are where we are.,* Certainly this book can be improved, and we are looking forward to receiving critical reviews of the book from students, teachers, and practitioners. We hope you will all gain as much knowledge, understanding and pleasure from reading this book as we have from writing it.

U Dinesh Kumar

Indian Institute of Management Calcutta, Kolkata, India

John Crocker

Data Systems and Solutions, UK

T Chitra

Ex Scientist, Defence Research and Development Organization, Bangalore, India

Haritha Saranga

Indian School of Business, Hyderabad, India

Acknowledgments

It gives us a great pleasure in thanking Dr Peter Hinds, CEO Karrus Limited UK and Mr H S Pannu, Chief Mechanical Engineer at South Eastern Railways, India for their assistance with some of the case studies of this book. Special thanks to Indian Institute of Management Calcutta and Indian School of Business, Hyderabad for providing the resources required for completing this book.

We would like to thank our families, partners, friends and colleagues without whose support and patience we would never have completed our task.

Finally we would like to thank Gary Folven and Carolyn Ford of Springer US for their continuous help and encouragement throughout the preparation of this book.

Chapter 1

RELIABILITY AND SIX SIGMA - INTRODUCTION

Every new body of discovery is mathematical in form, because there is no other guidance we can have.

Charles Darwin

1.1 INTRODUCTION

Every morning in Bombay (now Mumbai) about 5,000 dabbawallas (Tiffin carriers- Figure 1.1) collect about 200,000 lunch boxes from houses in various suburbs of Bombay, carry them in the suburban trains, and deliver them to various offices, colleges and schools around Bombay so that its citizens can eat fresh home made food. The customers are charged approximately $5 per month for this service. The most amazing fact about this massive logistics operation is that the dabbawallas almost never fail to deliver the lunch boxes to their rightful owners.

The Bombay dabbawallas is a six sigma organization (6.66 Sigma to be precise), that is, their failure rate (failure to deliver the lunch box to its rightful owner) is approximately 1 in 8 million. This high quality of service is achieved using simple processes, neither investing on infrastructure and management consultants nor reading books on strategy. There are no computer control systems to track the location of the lunch boxes during its two hour transit from Bombay houses through the suburban railways of Bombay to its owner; such as one you would normally come across with major courier companies like FedEx and UPS. Most members of the workforce of 5000 dabbawallas are illiterate. The entire logistics is coordinated by brain (or rather 5,000 brains) of these dabbawallas. Far away from Bombay, in the USA, many big companies spend millions of dollars to achieve Six Sigma quality, to the point where it has become almost an obsession. For example, 3M which started Six Sigma initiative within its organization in February 2001 completed about 11000 Six Sigma projects by August 2004. Companies like Motorola and General Electric attributed their success to their Six Sigma program.

Figure 1.1 Bombay Dabbawallas

Table 1-1. Sigma level of the US airline companies for baggage handling

Airline	Number of Mishandled Baggage per 1000	Sigma Level
American	4.60	4.10
Continental	4.29	4.13
Delta	4.11	4.14
Northwest	4.19	4.14
Southwest	4.77	4.09
Trans World	6.35	3.99
United	5.07	4.07

Just for comparison, we analyze the delivery quality achieved by Bombay dabbawallas against some of the leading airlines and their quality in delivering the checked-in baggage (Table 1.1, based on the data from Bowen *et al* 2002). It should be noted that the airline baggage handling system used by these airlines uses some of the most sophisticated technologies, but still as far as the quality of delivery is concerned, it has a lot to catch up and is far behind the dabbawallas. But, the good news is that the airline safety and reliability is much higher than Seven Sigma, which makes it one of the safest modes of transport. Thanks to Boeing and Airbus, the passengers are delivered to their destinations at a higher Sigma level quality.

Ever since the industrial revolution began about 3 centuries ago, customers have demanded better and cheaper products and services. Decision to procure a product depended on the quality and reliability of the product. Today's customers are no different. The only differences are that the companies have grown bigger, the products have become more sophisticated, complex and expensive and the market is more highly competitive. At the same time, customers have become more demanding, more mobile and more selective but less loyal to a product, service or a brand. Selection is likely to be based on perceptions and reputations as well as more tangible factors such as cost. Customers would like to possess a product which gives them superior performance at the lowest possible total cost of ownership. As in all forms of evolution, the *Red Queen Syndrome* is forever present: in business, as in all other things, you simply have to keep running faster and faster to stand still. No matter how good you make a product or service, it will never remain good in the eyes of the customer for very long.

1.2 WHAT IS SIX SIGMA?

Sigma (σ) is a Greek letter that has been used to represent standard deviation in statistics for many centuries. Standard deviation measures the spread of data from its mean value. Since the days of Deming and Juran, it is well accepted that variability within the process is one of the main causes for poor quality. The term Six Sigma was coined by William Smith, a reliability engineer at Motorola, and is a registered trade mark of Motorola. Bill Smith pioneered the concept of Six Sigma to deal with the high failure rate experienced by systems developed by Motorola. William Smith proposed Six Sigma as a goal to improve reliability and quality of the product. Until then, lower and upper control limits (LCL and UCL) for processes are fixed at 3σ deviations from the mean. Bill smith suggested that these limits (LCL and UCL) should be pushed to 6σ level. That is, the tolerance levels are set between μ-6σ and μ+6σ. This will force the designers to design their processes with minimum deviations (Figure 1.2).

In a three sigma process, the LCL and UCL are set at μ-3σ and μ+3σ respectively. The maximum variation, $\sigma_{max,3}$ allowed in a three sigma process is given by:

$$\sigma_{\max,3} = \left(\frac{UCL - LCL}{6}\right) \tag{1.1}$$

For example, if LCL = 3 and UCL = 9, then the maximum value of process variation, $\sigma_{max,3}$ is 1 (using equation 1.1). In a Six Sigma process, the lower and upper control limits are set at μ-6σ and μ+6σ respectively. The maximum variation, $\sigma_{max,6}$ allowed in a Six Sigma process is given by:

$$\sigma_{\max,6} = \left(\frac{UCL - LCL}{12}\right) \tag{1.2}$$

For same example with LCL = 3 and UCL = 9, the maximum value of process variation in a six sigma process from equation (1.2) is 1/2. That is, In a 3σ process, the maximum allowed process variation measured using standard deviation σ, is much higher (in fact double) compared to a Six Sigma process. Most importantly, changing a process from three sigma to Six Sigma has a significant impact on the number of defective parts per million (PPM).

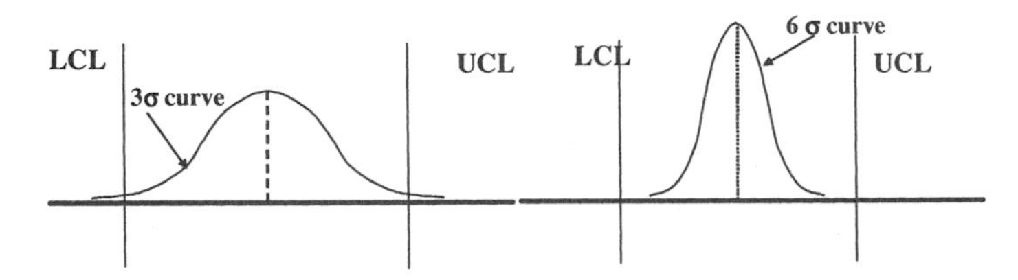

Figure 1-2. Comparison of 3s and 6s process

Six Sigma is a process improvement methodology, and over a period of time it has evolved into a management strategy which can be used to run an organization. Six Sigma is a customer focused quality improvement concept to achieve near perfect processes, products and services. Six Sigma is aimed at the near elimination of defects from every product, process and transaction. Six Sigma drives business processes to achieve a defect rate of less than 4 (3.4 to be precise) defects per million opportunities (DPMO). We would like to point out that having specification limits of six standard deviations (σ) from the mean of a normal distribution will not result in 3.4 defects per million opportunities. The number 3.4 is derived by assuming that the process mean itself may drift over a long period of time by 1.5 standard deviations (Figure 1.3). Researchers at Motorola found that the processes can shift by up to 1.5 standard deviations over a long period of time. Under this assumption (and belief that this in fact is true!), the area under the normal distribution beyond 4.5 σ is approximately 3.4 /1,000,000. That is 3.4 out of million opportunities lie outside the tolerance level. So, what should have been called a 4.5 σ is termed as 6 σ due to the belief that the process itself may shift by 1.5 σ in the long run, which may be true in some cases, but definitely not all.

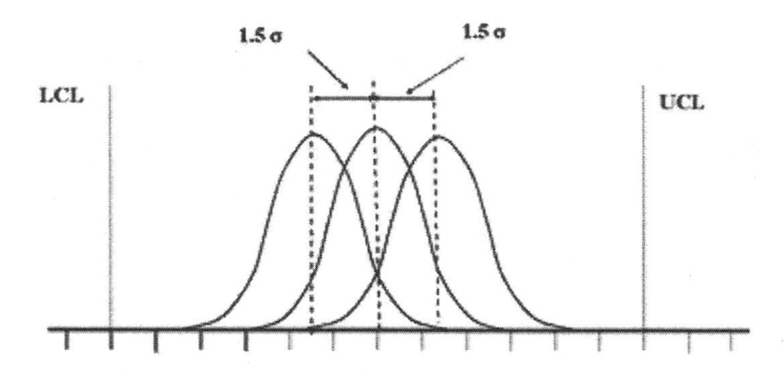

Figure 1-3. Process shift by 1.5 σ

The assumption that the process mean may drift by 1.5 σ about the average has received considerable attention and criticism in the literature. One reason for the shift in the process mean is due to the learning effect. For example consider an automobile company manufacturing cars. Fuel efficiency of the car is one of the important parameters of interest to customers. One can expect an improvement in the fuel efficiency of the car over a period of time due to learning that happens within the company once the data about the processes become available leading to the improvement in some of the processes. Suppose a car model is released with a mileage of 20 kilometers per litre, with standard deviation of 2 Kms. Over a period of time, the mileage of cars developed subsequent to the existing model is likely to increase due to improvement in the design and manufacturing processes. On the other hand, the mileage of a car is likely to decrease over a period of time due to wear out factors. Techniques such as relevant condition indicator, RCI, (see Chapter 6 for more details) can be used to estimate the actual drift in the process mean.

In the literature, there exist many definitions for Six Sigma. In this book, we define "Six Sigma" as follows:

Six Sigma is a management strategy which provides a roadmap to continuously improve business processes to eliminate defects in products, processes and service.

During the initials days of Six Sigma, the main focus was on reducing scrap and rework in manufacturing and thus reduce the manufacturing cost. However, over the last decade, Six Sigma has evolved into a management strategy which can be applied to any business that is looking for performance improvement. In the mid-1990s, Jack Welch, the then Chief Executive Office of General Electric (GE) proclaimed that GE was making great savings through Six Sigma strategy. During the first five years of the program, the company increased its productivity by over 266% and improved operating margins from 14.4% to 18.4%. Other major American firms then became Six Sigma devotees. Impressive business performance achieved by companies like Motorola, General Electric, Allied Signal, Black and Decker, Honeywell, ABB and Bombardier can be attributed to Six Sigma strategy. In fact, their annual reports clearly attribute their success to Six Sigma strategies.

1.3 WHAT IS NEW IN SIX SIGMA?

Six Sigma is an extension of other quality initiatives, starting from Deming's statistical quality control. Most of the management strategies on quality initiatives are customer focused and Six Sigma is also focused around meeting the customer requirements as the main objective. In fact, all the quality initiatives developed in the 20th century are customer focused. Ogilvy once said, 'Customer is your wife'. This analogy reminds us of a joke. One Yankee was walking on a Californian beach and hits upon a cola bottle. Having found the wonder drink, he opens the bottle and instead of drink, finds a genie coming out. Genie says, 'Thank you Yankee for releasing me from the Cola bottle, however, we Genies are currently cutting costs, so instead of regular three wishes, you get only one wish, so go on, make a wish'. Yankee thinks for some time and says, 'Genie, I have heard a lot about Hawaii, and it is my lifetime ambition, to visit Hawaii at least once. However, I have a problem; I am very scared of flying. So, can you please build a bridge between California and Hawaii, so that I can drive my car?' But Genie tells Yankee that is not possible, since it will require a huge amount of concrete and steel, especially when they are cutting costs, it is not feasible, so asks him for some other wish. 'Well Genie, I will ask for a much simpler one, I love my wife very much, I care for her a lot. But she doesn't think that I love her. She complains all the time that I don't care for her, and I spend more time with the TV remote than with her. Can you tell me how to keep my wife happy? I would really appreciate that'. Genie thinks for some time and asks the Yankee, 'now about that bridge, are you looking for two lanes or four lanes?' Unfortunately, businesses are expected to understand the real customer (who most often is not happy with you, just like your wife) need, which of course is the most difficult task in any product development. Once the customer requirements are understood, the business should ensure that they meet all requirements.

The main difference between other quality initiatives and Six Sigma is the disciplined quantitative approach used by Six Sigma for the process improvement. Six Sigma uses five macro phases for improvement: Define, Measure, Analyze, Improve and Control (DMAIC). DMAIC cycle is a standardized methodology used by Six Sigma to achieve process improvement. However, the DMAIC cycle itself has a lot of similarities with Deming's "Plan-Do-Check-Act" (PDCA) cycle. Where Six Sigma differs is that it provides a well defined target for quality, which is not more than 3.4 defects per million opportunities. Another important difference between Six Sigma and Total Quality Management (TQM) is that Six Sigma is mostly a business results oriented model compared to a return on investment orientation of Total Quality Management (Bertels, 2003). Six Sigma can either be used as an operational strategy to reduce the number of defects or as a business strategy to improve business processes and evolve new business models. Design for Six Sigma (DFSS) concept uses Six Sigma to as a strategy to design and develop products compared to the traditional Six Sigma which aims to reduce the defects.

A survey conducted by DynCorp has revealed that the concept of Six Sigma is rated highly compared to many other process improvement concepts. Table 1.2 shows rating of Six Sigma and other process improvement techniques based on the survey conducted by the DynCorp to determine the most successful process improvement tool.

Table 1-2. Rating of Six Sigma and other process improvement techniques

Process Improvement Tool	Impact
Six Sigma	53.6%
Process Mapping	35.3%
Root Cause Analysis	33.5%
Cause and Effect Analysis	31.3%
ISO 9001	21.0%
Statistical Process Control	20.1%
Total Quality Management	10.3%
Malcolm Baldridge Criteria	9.8%
Knowledge Management	5.8%

1.4 QUALITY AND SIX SIGMA LEVEL

The quality of a manufacturing process can be measured using sigma levels (also called sigma quality level, discussed in detail in Chapter 3). Table 1.3 gives the number of failures/defects per million opportunities for various sigma levels. These were derived assuming that the number of defects in a sample is normally distributed and the process mean itself can shift up to 1.5 σ. Table 1.4 shows the number of defects per million opportunities without shift in the process mean.

Table 1-3. Sigma level and Defects Per Million Opportunities with 1.5σ shift

Sigma Level	DPMO
1	697,672
2	308,770
3	66,811
4	6,210
5	233
6	3.4

Table 1-4. Sigma level and Defects Per Million Opportunities without shift in the process mean

Sigma Level	DPMO
1	317,300
2	45,500
3	2700
4	63
5	0.57
6	0.002

Sigma level can be used as a measure for benchmarking purposes and to discriminate best processes from weak processes.

1.5 RELIABILITY AND SIX SIGMA

The concept of Six Sigma evolved while trying to solve a reliability problem. Bill Smith, while working at Motorola's Communications Division, noticed a high failure rate on their products and suggested a higher level of internal quality and set Six Sigma as a quality target. He believed that Six Sigma quality target would improve the reliability as measured in mean time to failure and quality as measured in defect rates.

Initially Six Sigma was developed to improve the manufacturing processes; however, the concept can be used to improve any process, not just manufacturing processes. It is important to understand that a manufacturer cannot satisfy customers by providing defect free products alone. One of the important design characteristics of any product is the reliability of that product, as it affects the utility of the product and the total cost of ownership of that product. Table 1.5 shows predicted and observed mean time between failures (MTBF) of SINCGARS (Single Channel Ground to Air Radio System) and the corresponding sigma level (see chapter 3 for calculation of sigma level, in case of vendor G the sigma level is ∞, theoretically Sigma level can be ∞). The reliability requirement by the Army was 1250 hours with 80% confidence. After collecting data over a long period, the Army realized that there was a huge difference between the predicted MTBF (predicted using MIL-HDBK-217) and the observed MTBF. A defective product can be easily identified while commissioning the product, whereas, a product with poor reliability hurts the customer more. It increases the number of maintenance tasks, and logistics requirements and thus ends up with a higher total cost of ownership.

Table 1-5. SINGCARS predicted, observed MTBF and Six Sigma

Vendor	Predicted MTBF	Observed MTBF	Sigma level
A	7247	1160	4.684902
B	5765	74	3.716165
C	3500	624	4.507558
D	2500	2174	5.346291
E	2500	51	3.57039
F	2000	1056	4.821975
G	1600	3612	∞
H	1400	98	3.845931
I	1000	472	4.556798

The primary goal in any business is customer satisfaction (if not customer delight). This would mean that every business should know what the customers really want. In some cases, this may be a tricky situation. Techniques such as Quality Function Deployment (QFD) may help an organization to capture what customers really want. But, the problem is that the customer requirements are likely to change every day. Achieving Six Sigma quality means designing the process with a defect rate of 3.4 defects per million opportunities; however, achieving Six Sigma quality at the production stage alone cannot guarantee success for an organization. Product reliability, without any doubt, is one of the most important requirements of the customer. No customer would like to see his or her product failing within a short period of purchase, even if the product is still under warranty. The effectiveness of an organization should be measured on the basis of the production effectiveness (or manufacturing effectiveness), product effectiveness and market effectiveness.

Production effectiveness refers to the quality of manufacturing and can be measured using Six Sigma.

Product Effectiveness refers to the quality of design as measured in terms of reliability of that product - effectiveness in meeting customer requirements related to performance.

Market effectiveness refers to the quality of the product features that provides the product with a competitive edge in the market.

In this book, our focus is on product effectiveness (reliability) from a customer perspective and production effectiveness from a manufacturer's perspective. To achieve market effectiveness, it is important to ensure that the product is effective as measured in terms of reliability and ability to meet customer requirements such as performance, maintainability, supportability, cost of ownership etc as well as the effectiveness of the production process which can be measured using Six Sigma. For this reason, it is important to consider the concepts of Six Sigma and Reliability together instead of two isolated concepts. What hurts the customer more is the breakdown of the product while it is in use rather than a product which is dead on arrival.

Chapter 2

RELIABILITY AND SIX SIGMA – PROBABILITIC MODELS

Is man one of God's blunders or is God one of man's blunders?
- Friedrich Nietzsche

2.1 INTRODUCTION

Development of any product involves designing processes that can result in the desired product. Successful development of a product depends on the effectiveness of the underlying processes. Naturally, process variability is a concern for any product development team, since it affects the reliability and quality of the product. To manage the processes, we have to measure it, especially the variations and the uncertainties associated with them. Apart from various other factors, market success of a product would depend on how the company manages to meet the customer requirements under uncertainties. For example, customers may be happy with a courier company which delivers documents with mean duration of two days and standard deviation of one day compared to another company which delivers the documents with mean duration of one day with standard deviation of two days. In the first case, the probability that a document will not be delivered within 4 days is approximately 2.27% (assuming normal distribution), whereas the same probability is 6.67% in case of second company, although its average time to deliver the document is one day. Similarly, a customer purchasing a product such as consumer durables would expect the product to function properly to begin with and maintain the functionality for a minimum duration. This 'minimum duration' is subjective and would vary from product to product. Here, there are two uncertainties; one corresponds to the probability that a newly purchased product will function, and two the product will maintain the functionality for a minimum period of time. The former corresponds to Six Sigma and the latter is related to the reliability of the product.

Probability theory is the fundamental building block for both reliability theory and Six Sigma. In this chapter we introduce the basic concepts in probability theory which are essential to understand the rest of the book. This chapter is not intended for a rigorous treatment of all-relevant theorems and proofs in probability theory, but to provide an understanding of the main concepts in probability theory that can be applied to problems in reliability and Six Sigma.

2.2 PROBABILITY TERMINOLOGY

In this section we introduce various terminologies used in probability that are essential for understanding the rudiments of probability theory. To facilitate the discussion some relevant terms and their definitions are introduced.

Experiment

An experiment is a well-defined act or process that leads to a single well-defined outcome. Figure 2.1 illustrates the concept of random experiments. Every experiment must:

1. Be capable of being described, so that the observer knows when it occurs.
2. Have one and only one outcome.

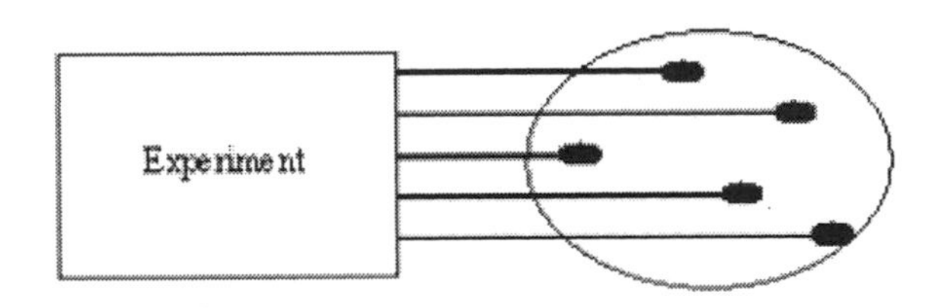

Figure 2-1. Graphical Representation of an Experiment and its outcomes

Elementary event

An elementary event is every separate outcome of an experiment, also known as a sample point.

From the definition of an experiment, it is possible to conclude that the total number of elementary events is equal to the total number of possible outcomes.

Sample space

The set of all possible distinct outcomes for an experiment is called the sample space for that experiment.

Usually, the symbol S is used to represent the *sample space*, and small letters, a, b, c,.., for elementary events that are possible outcomes of the experiment under consideration. The set S may contain either a finite or infinite number of elementary events. Figure 2.2 is a graphical presentation of the sample space.

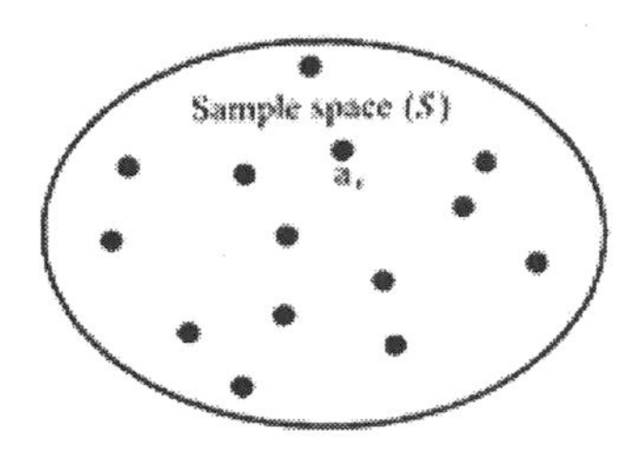

Figure 2-2. Graphical Presentation of the Sample Space

Event

Event is a subset of the sample space, that is, a collection of elementary events.

Capital letters A, B, C,... are usually used for denoting events. For example, if the experiment performed is measuring the speed of passing cars at a specific road junction, then the elementary event is the speed measured, whereas the sample space consists of all the different speeds one might possibly record. All speed events could be classified in to, say, four different speed groups: A (less than 30 km/h), B (between 30 and 50 km/h), C (between 50 and 70 km/h) and D (above 70 km/h). If the measured speed of the passing car is, say 35 km/h, then the event B is said to have occurred.

2.3 ELEMENTARY THEORY OF PROBABILITY

The theory of probability is developed from axioms proposed by the Russian mathematician *Kolmogrov*. All other rules and relations in probability theory are derived from axioms of probability.

2.3.1 Axioms of probability

In cases where the outcome of an experiment is uncertain, it is necessary to assign some measure that will indicate the chances of occurrence of a particular event. Such a measure of events is called the *probability of the event* and symbolized by *P(.)*, (*P(A)* denotes the probability of occurrence of event *A*). The function which associates each event *A* in the sample space *S*, with the probability measure *P(A)*, is called the *probability function* - the probability of that event. A graphical representation of the probability function is given in Figure 2.3.

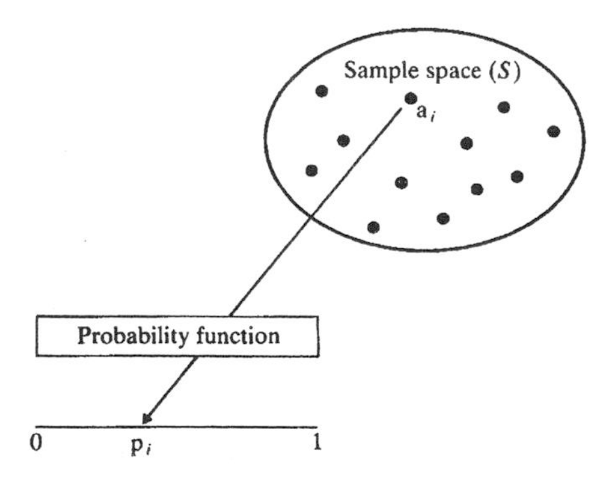

Figure 2-3. Graphical representation of probability function

Formally, the probability function is defined as:

A function which associates with each event A, a real number, P(A), the probability of event A, such that the following axioms are true:

1. $P(A) > 0$ for every event A,
2. $P(S) = 1$, (probability of the sample space is equal to one)
3. The probability of the union of mutually exclusive events is the sum of their probabilities, that is

$$P(A_1 \cup A_2 ... \cup A_n) = P(A_1) + P(A_2) + ... + P(A_n)$$

In essence, this definition states that each event A is paired with a non-negative number, probability P(A), and that the probability of the sure event S, or P(S), is always 1. Furthermore, if A_1 and A_2 are any two mutually exclusive events (that is, the occurrence of one event implies the non-occurrence of the other) in the sample space, then the probability of their union $P(A_1 \cup A_2)$, is simply the sum of their two probabilities, $P(A_1) + P(A_2)$.

2.3.2 Rules of probability

The following elementary rules of probability are directly deduced from the original three axioms, using set theory:

a) For any event A, the probability of the complementary event, denoted by A', is given by

$$P(A') = 1 - P(A) \tag{2.1}$$

b) The probability of any event must lie between zero and one:

$$0 \leq P(A) \leq 1 \tag{2.2}$$

c) The probability of an empty or impossible event, ϕ, is zero.

$$P(\phi) = 0 \tag{2.3}$$

d) If occurrence of an event A implies occurrence of an event B, so that the event class A is a subset of event class B, then the probability of A is less than or equal to the probability of B:

$$P(A) \leq P(B) \tag{2.4}$$

e) In order to find the probability that A or B or both occur, the probability of A, the probability of B, and also the probability that both occur must be known, thus:

$$P(A \cup B) = P(A) + P(B) - P(A \cap B) \tag{2.5}$$

f) If A and B are mutually exclusive events, so that $P(A \cap B) = 0$, then

$$P(A \cup B) = P(A) + P(B) \tag{2.6}$$

g) If n events form a partition of the sample space S, then their probabilities must add up to one:

$$P(A_1) + P(A_2) + ... + P(A_n) = \sum_{i=1}^{n} P(A_i) = 1 \tag{2.7}$$

2.3.3 Joint event

Any event that is an intersection of two or more events is called a joint event.

There is nothing to restrict any given elementary event from the sample space from qualifying for two or more events, provided that those events are not mutually exclusive. Thus, given the event *A* and the event *B*, the joint event is $A \cap B$. Since a member of $A \cap B$ must be a member of set *A*, and also of set *B*, both *A* and *B* events occur when $A \cap B$ occurs. Provided that the elements of set *S* are all equally likely to occur, the probability of the joint event could be found in the following way:

$$P(A \cap B) = \frac{\text{number of elementary events in A} \cap \text{B}}{\text{total number of elementary events}}$$

2.3.4 Conditional probability

If A and B are events in a sample space which consists of a finite number of elementary events, the conditional probability of the event B given that the event A has already occurred, denoted by $P(B|A)$, is defined as:

$$P(B|A) = \frac{P(A \cap B)}{P(A)}, \qquad P(A) > 0 \tag{2.8}$$

The conditional probability symbol, $P(B|A)$, is read as the probability of occurrence of event *B* given that the event *A* has occurred. It is necessary to satisfy the condition that $P(A)>0$, because it does not make sense to consider the probability of *B* given *A* if event *A* is impossible. For any two events *A* and *B*, there are two conditional probabilities that may be calculated:

$$P(B|A) = \frac{P(A \cap B)}{P(A)} \qquad \textit{and} \qquad P(A|B) = \frac{P(A \cap B)}{P(B)}$$

(The probability of B, given A) (The probability of A, given B)

One of the important applications of conditional probability is due to Bayes theorem, which can be stated as follows:

If $(A_1, A_2, \ldots, A_N)$ represents the partition of the sample space (*N* mutually exclusive events), and if *B* is subset of $(A_1 \cup A_2 \cup \ldots \cup A_N)$, as illustrated in Figure 2.4, then

$$P(A_i|B) = \frac{P(B|A_i)P(A_i)}{P(B|A_1)P(A_1)+\ldots+P(B|A_i)P(A_i)+\ldots+P(B|A_N)P(A_N)} \tag{2.9}$$

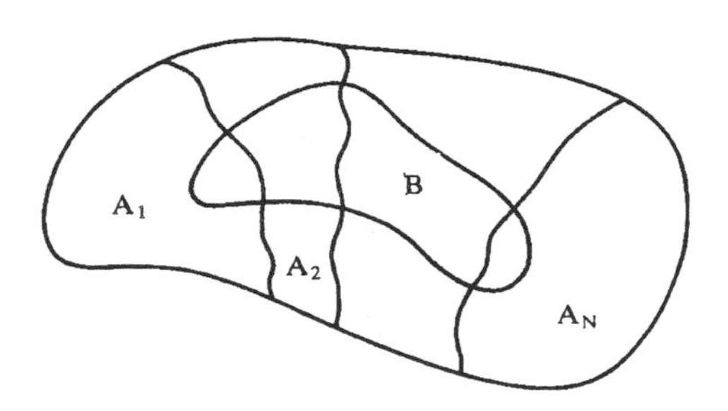

Figure 2-4. Graphical Presentation of the Bayes Theorem

2.4 PROBABILITY DISTRIBUTION

Consider the set of events $A_1, A_2, \ldots, A_n$, and suppose that they form a partition of the sample space *S*. That is, they are mutually exclusive and exhaustive. Then the corresponding set of probabilities, $P(A_1), P(A_2), \ldots, P(A_n)$, is a probability distribution. An illustrative presentation of the concept of probability distribution is shown in Figure 2.5.

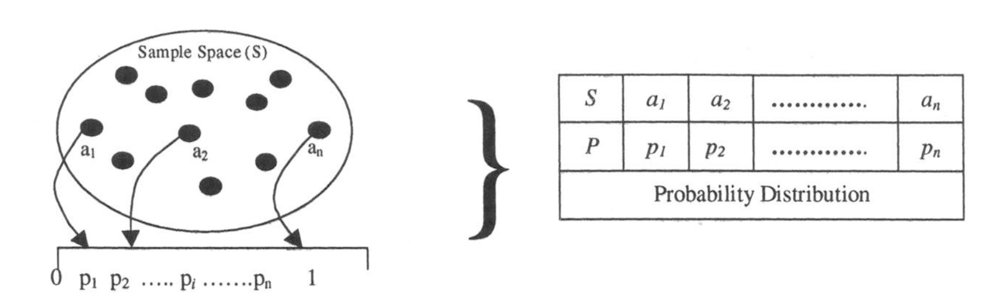

Figure 2-5. Graphical representation of Probability Distribution

2.5 RANDOM VARIABLE

A function that assigns a number (usually a real number) to each sample point in the sample space S is a random variable.

Outcomes of experiments may be expressed either in numerical or non-numerical terms. However, In order to compare and analyze them it is convenient to use real numbers. So, for practical applications, it is necessary to assign a numerical value to each possible elementary event in a sample space S. Even if the elementary events themselves are already expressed in terms of numbers, it is possible to reassign a unique real number to each elementary event. The function that achieves this is known as *the random variable*. In other words, a random variable is a real-valued function defined on a sample space. Usually it is denoted with capital letters, such as X, Y and Z, whereas small letters, such as x, y, z, a, b, c, and so on, are used to denote particular values of a random variables, see Figure 2.6.

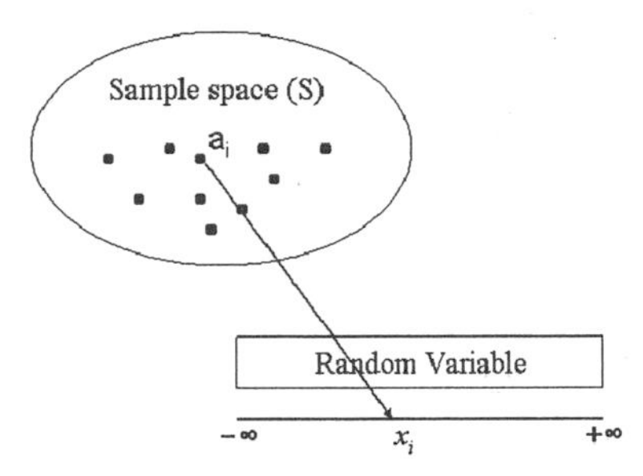

Figure 2-6. Graphical Representation of Random Variable

If X is a random variable and r is a fixed real number, it is possible to define the event A to be the subset of S consisting of all sample points 'a' to which the random variable X assigns the number r, $A = (a : X(a) = r)$. On the other hand, the event A has a probability $p = P(A)$. The symbol p can be interpreted, generally, as the probability that the random variable X takes value r, $p = P(X = r)$. Thus, the symbol $P(X = r)$ represents the probability function of a random variable. Therefore, by using the random variable it is possible to assign probabilities to real numbers, although the original probabilities were only defined for events of the set S, as shown in Figure 2.7.

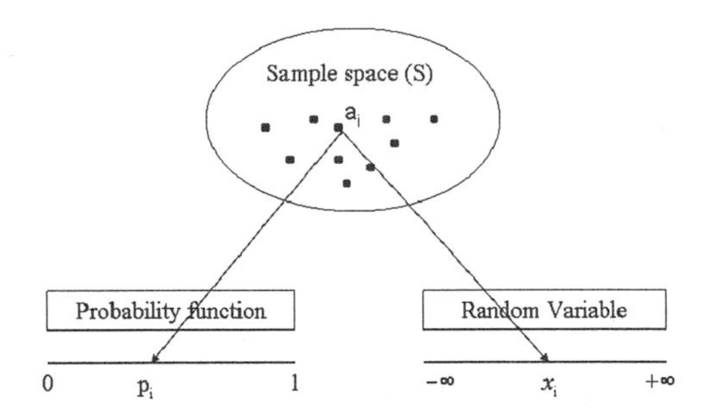

Figure 2-7. Relationship between probability function and a random variable

The probability that the random variable X, takes value less than or equal to certain value 'x', is called the *cumulative distribution function, F(t).* That is,

$$P[X \leq x] = F(x)$$

2.5.1 Types of random variables

Depending on the values assumed, random variables can be classified as discrete or continuous. The main characteristics, similarities and differences for both types are briefly described below.

Discrete random variables

If the random variable X can assume only a particular finite or countably infinite set of values, it is said to be a discrete random variable.

There are very many situations where the random variable X can assume only a *finite* or *countably infinite* set of values; that is, the possible values of X are finite in number or they are infinite in number but can be put in a one-to-one correspondence with a set of real number.

Continuous random variables

If the random variable X can assume any value from a finite or an infinite set of values, it is said to be a continuous random variable.

Let us consider an experiment, which consists of recording the temperature of a cooling liquid of an engine in the area of the thermostat at a given time. Suppose that we can measure the temperature exactly, which means that our measuring device allows us to record the temperature to any number of decimal points. If X is the temperature reading, it is not possible for us to specify a finite or countably infinite set of values. For example, if one of the finite set of values is 75.965, we can determine values 75.9651, 75.9652, and so on, which are also possible values of X. What is being demonstrated here is that the possible values of X consist of the set of real numbers, a set which contains an infinite (and uncountable) number of values. Continuous random variables have enormous utility in reliability since the random variables time to failure is a continuous random variable.

2.6 THE PROBABILITY DISTRIBUTION OF A RANDOM VARIABLE

Taking into account the concept of the probability distribution and the concept of a random variable, it could be said that the probability distribution of a random variable is a set of pairs:

$$\{r_i, P(X = r_i),\ i = 1,n\} \tag{2.10}$$

Figure 2.8 shows the relationship between a random variable and its probability distribution. The easiest way to present this set is to make a list of all its members. If the number of possible values is small, it is easy to specify a probability distribution. On the other hand, if there are a large number of possible values, a listing may become very difficult. In the extreme case where we have an infinite number of possible values (for example, all real numbers between zero and one), it is clearly impossible to make a listing. Fortunately, there are other methods that could be used for specifying a probability distribution of a random variable:

1. Functional method, where a specific mathematical function exists from which the probability of random variable taking any value or interval of values can be calculated.
2. Parametric method, where the entire distribution is represented through one or more parameters known as summary measures.

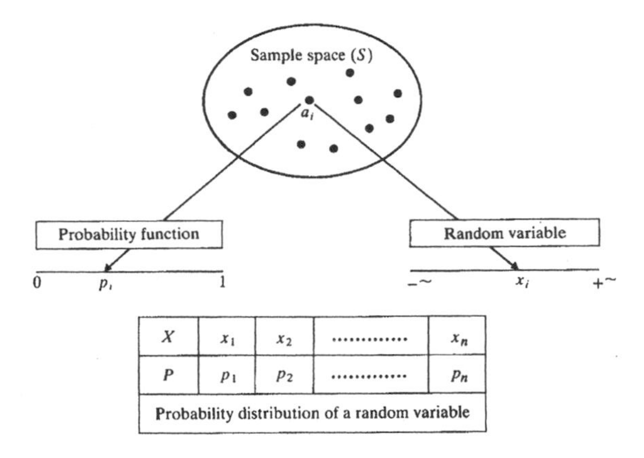

X	x_1	x_2	....	x_n
P	p_1	p_2	....	p_n

Figure 2-8. Probability Distribution of a Random Variable

2.6.1 Functional method

By definition, a function is a relation where each member of the domain is paired with one member of the range. In this particular case, the relation between numerical values that random variables assume and their probabilities will be considered. The most frequently used functions for the description of probability distribution of a random variable are the probability mass function, the probability density function, and the cumulative distribution function. Each of these will be analyzed and defined in the remainder of this chapter.

Probability mass function

This function is related to a discrete random variable and it represents the probability that the discrete random variable, *X*, will take one specific value x_i, $p_i = P(X = x_i)$. Thus, a probability mass function, which is usually denoted as $PMF(.)$, places a mass of probability p_i at the point x_i on the X-axis.

Given that a discrete random variable takes on only *n* different values, say $a_1, a_2, \ldots, a_n$, the corresponding $PMF(.)$ must satisfy the following two conditions:

1. $P(X = a_i) \geq 0 \qquad for\ i = 1,2,\ldots,n$

2. $\sum_{i=1}^{n} P(X = a_i) = 1 \qquad (2.11)$

In practice this means that the probability of each value that *X* can take must be non-negative and the sum of the probabilities must be 1. Thus, a probability distribution can be represented by a pair of values (a_i, p_i), where $i = 1,2,\ldots,n$, as shown in Figure 2.9. The advantage of such a graph over a listing is the ease of comprehension and a better provision of a notion for the nature of the probability distribution.

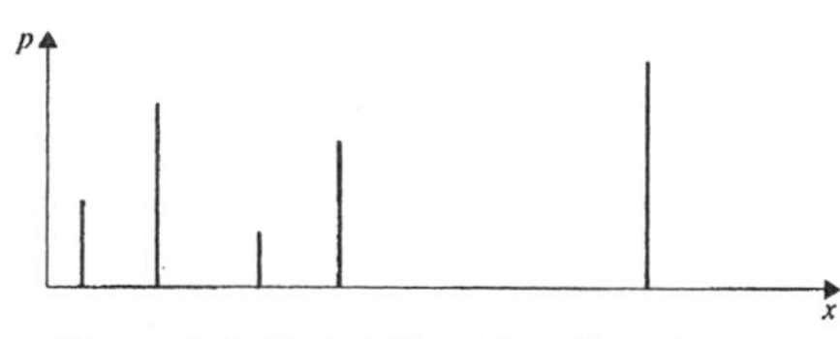

Figure 2-9. Probability Mass Function

Probability density function

In the previous section discrete random variables were discussed in terms of probabilities *P*(*X* =*x*), the probability that the random variables take an *exact* value. However, consider the example of an infinite set. For a specific type of car, the volume of the fuel in the fuel tank is measured with only some degree of accuracy. What is the probability that a car selected at random will have *exactly* 16 litres of fuel? This could be considered as an event that is defined by the interval of values between, say 15.5 and 16.5, or 15.75 and 16.25, or any other interval $16 \pm 0.1i$, where i is very small, but not exactly zero. Since the smaller the interval, the smaller the probability, the probability of exactly 16 litres is, in effect, zero.

In general, for continuous random variables, the occurrence of any exact value of X may be regarded as having zero probability.

The Probability Density Function, $f(x)$, which represents the probability that the random variable will take values within the interval $x \le X \le x + \Delta(x)$, where $\Delta(x)$ approaches zero, is defined as:

$$f(x) = \lim_{\Delta(x) \to 0} \frac{P(x \le X \le x + \Delta(x))}{\Delta x} \qquad (2.12)$$

As a consequence, the probabilities of a continuous random variable can be discussed only for *intervals* that the random variable *X* can take. Thus, instead of the probability that *X* takes on a specific value, say *'a'*, we deal with the so-called *probability density* of *X* at *'a'*, symbolized by $f(a)$. In general, the probability distribution of a continuous random variable can be represented by its *Probability Density Function, PDF*, which is defined in the following way:

$$P(a \le X \le b) = \int_a^b f(x)dx \qquad (2.13)$$

A fully defined probability density function must satisfy the following two requirements:

$$f(x) \ge 0 \qquad \textit{for all } x$$

$$\int_{-\infty}^{+\infty} f(x)dx = 1$$

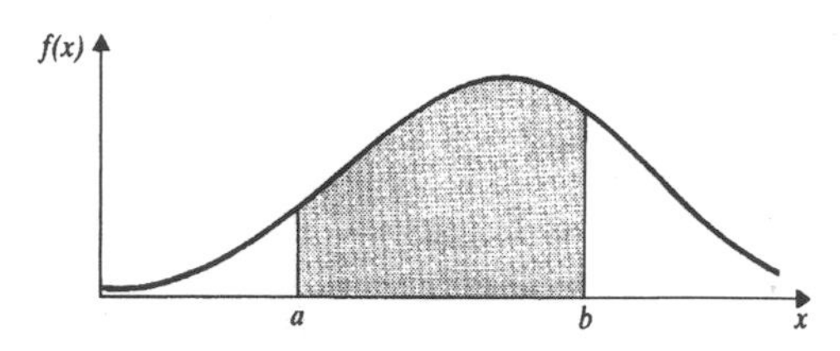

Figure 2-10. Probability Density Function for a Hypothetical Distribution

The *PDF* is always represented as a smooth curve drawn above the horizontal axis, which represents the possible values of the random variable X. A curve for a hypothetical distribution is shown in Figure 2.10 where the two points a and b on the horizontal axis represent limits which define an interval. The shaded portion between *'a'* and *'b'* represents the probability that X takes a value between the limits *'a'* and *'b'*.

Cumulative distribution function

The probability that a random variable X takes a value at or below a given number 'a' is often written as:

$$F(a) = P(X \leq a) \tag{2.14}$$

The symbol $F(a)$ denotes the particular probability for the interval $X \leq a$. This function is called the *Cumulative Distribution Function, CDF*, and it must satisfy certain mathematical properties, the most important of which are:

1. $0 \leq F(x) \leq 1$
2. $if \ \ a < b, \qquad F(a) \leq F(b)$

   $F(\infty) = 1 \qquad and \qquad F(-\infty) = 0$

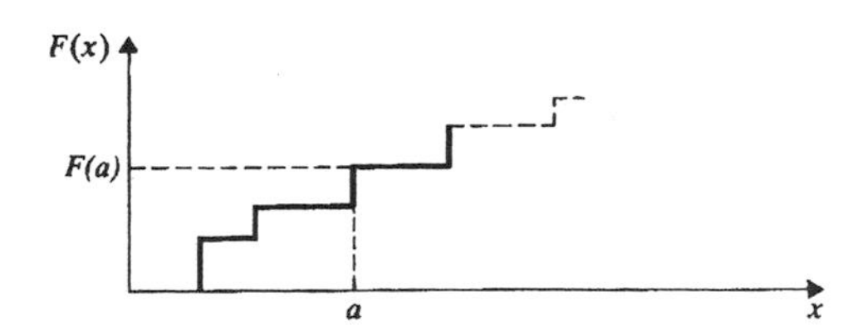

Figure 2-11. Cumulative Distribution Function for Discrete Variable

In general, the symbol $F(x)$ can be used to represent the cumulative probability that X is less than or equal to x. For the discrete random variable, it is defined as:

$$F(a) = \sum_{i=1}^{n} P(X = x_i) \tag{2.15}$$

Whereas in the case of continuous random variables it will take the following form:

$$F(a) = \int_{-\infty}^{a} f(x)dx \tag{2.16}$$

Hypothetical cumulative distribution functions for both types of random variable are given in Figures 2.11 and 2.12.

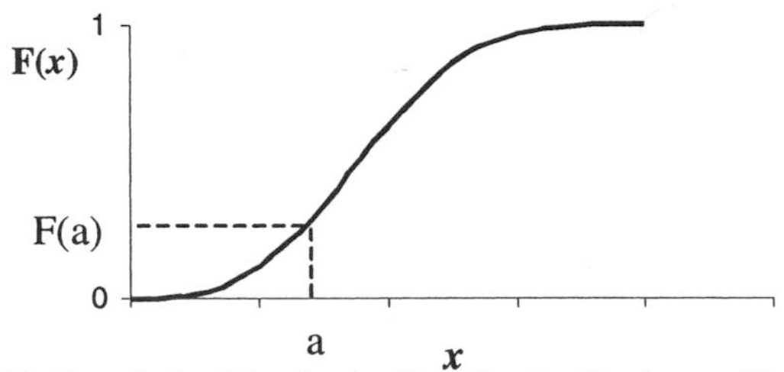

Figure 2-12. Cumulative Distribution Function for Continuous Variable

2.6.2 Parametric method

In some situations it is easier and more efficient to look only at certain characteristics of distributions rather than to attempt to specify the distribution as a whole. Such characteristics summarize and numerically describe certain features for the entire distribution. Two general groups of such characteristics applicable to any type of distribution are:

a) *Measures of central tendency* (or location) which indicate the typical or the average value of the random variable.
b) *Measures of dispersion* (or variability) which show the spread of the difference among the possible values of the random variable.

In many cases, it is possible to adequately describe a probability distribution with a few measures of this kind. It should be remembered, however, that these measures serve only to summarize some important features of the probability distribution. In general, they do not completely describe the entire distribution.

One of the most common and useful summary measures of a probability distribution is *the expectation* of a random variable, *E(X)*. It is a unique value that indicates a location for the distribution as a whole (In physical science, expected value actually represents the centre of gravity). The concept of expectation plays an important role not only as a useful measure, but also as a central concept within the theory of probability and statistics.

If a random variable, say *X*, is discrete, then its expectation is defined as:

$$E(X) = \sum_{x} x \times P(X = x) \qquad (2.17)$$

Where the summation is over all values the variable *X* can take. If the random variable is continuous, then the expectation is defined as:

$$E(X) = \int_{-\infty}^{+\infty} x \times f(x) dx \qquad (2.18)$$

For a continuous random variable the expectation is also defined as:

$$E(X) = \int_{-\infty}^{+\infty} [1 - F(x)] dx \qquad (2.19)$$

If *c* is a constant, then

$$E(cX) = c \times E(X) \qquad (2.20)$$

Also, for any two random variables *X* and *Y*,

$$E(X + Y) = E(X) + E(Y)$$

2.6.2.1 Measures of central tendency

The most frequently used measures are mean, median and mode.

The mean of a random variable is simply the expectation of the random variable under consideration. Thus, for the random variable, *X*, the mean value is defined as:

$$Mean = E(X) \tag{2.21}$$

The median, is defined as the value of *X* which is midway (in terms of probability) between the smallest possible value and the largest possible value. The median is the point, which divides the total area under the *PDF* into two equal parts. In other words, the probability that *X* is less than the median is $1/2$, and the probability that *X* is greater than the median is also $1/2$. Thus, if $P(X \leq a) = 0.50$ and $P(X \geq a) = 0.50$ then *'a'* is the *median* of the distribution of *X*. In the continuous case, this can be expressed as:

$$\int_{-\infty}^{a} f(x)dx = \int_{a}^{+\infty} f(x)dx = 0.50 \tag{2.22}$$

The mode, is defined as the value of *X* at which the *PDF* of *X* reaches its highest point. If a graph of the *PMF (PDF)*, or a listing of possible values of *X* along with their probabilities is available, determination of the mode is quite simple.

A central tendency parameter, whether it is mode, median, mean, or any other measure, summarizes only a certain aspect of a distribution. It is easy to find two distributions which have the same mean but are not similar in any other respect.

2.6.2.2 Measures of dispersion

The mean is a good indication of the location of a random variable, but it is possible that *no single value of the random variable may match with mean*. A deviation from the mean, D, expresses the measure of error made by using the mean as measure of the random variable:

$$D = x - M$$

Where, *x*, is a possible value of the random variable, *X*. The deviation can be taken from other measures of central tendency as well, such as the median or mode. It is quite obvious that the larger such deviations are from a measure of central tendency, the more the individual values differ from each other, and the more apparent the spread within the distribution becomes. Consequently, it is necessary to find a measure that will reflect the spread, or variability, of individual values.

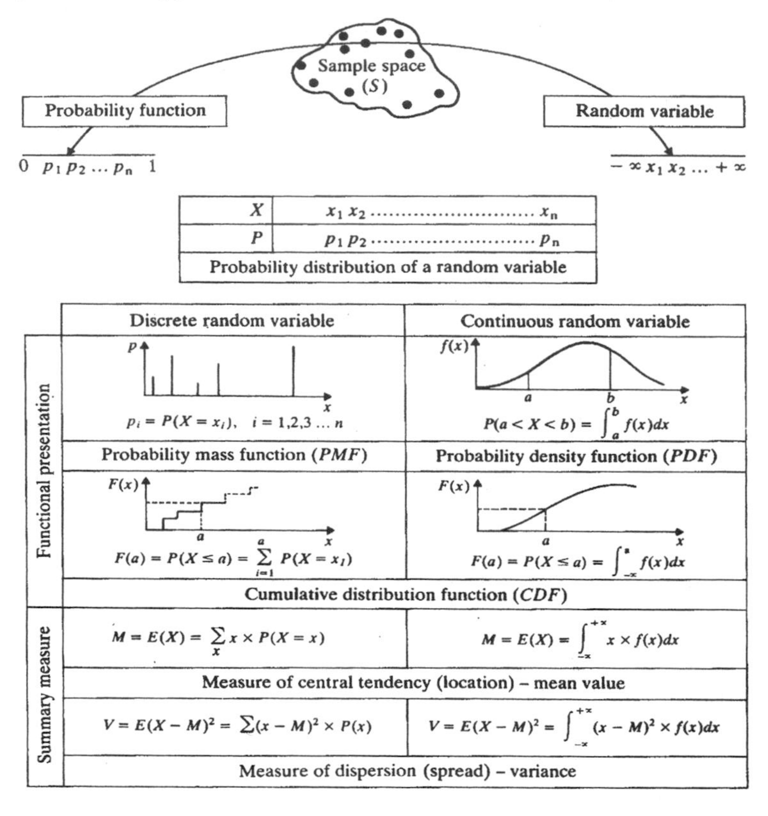

Figure 2-13. Probability System for Continuous Random Variable

The expectation of the deviation about the mean as a measure of variability, $E(X - M)$, will not work because the expected deviation from the mean must be zero for obvious reasons. The solution is to find the *square* of each deviation from the mean, and then to find the expectation of the squared deviation. This characteristic is known as a *variance of the distribution*, *V*, thus:

$$V(X) = E(X - Mean)^2 = \sum (X - Mean)^2 \times P(x) \qquad \text{if X is discrete} \qquad (2.23)$$

$$V(X) = E(X - Mean)^2 = \int_{-\infty}^{+\infty} (X - Mean)^2 \times f(x)dx \quad \text{if X is continuous} \qquad (2.24)$$

The positive square root of the variance of a distribution is called the *Standard Deviation, SD* (usually denoted using σ).

$$SD = \sigma = \sqrt{V(X)} \qquad (2.25)$$

Probability distributions can be analyzed in greater depth by introducing other summary measures, known as *moments*. Very simply these are expectations of different powers of the random variable. More information about them can be found in texts on probability.

2.6.2.3 Variability

The standard deviation is a measure that shows how closely the values of random variables are concentrated around the mean. Sometimes it is difficult to use only knowledge of the standard deviation, to decide whether the dispersion is considerably large or small, because this will depend on the mean value. In this case the parameter known as coefficient of variation, CV_X, defined in equation (2.26) can be used.

$$CV_X = \frac{SD}{M} \qquad (2.26)$$

Coefficient of variation is very useful because it gives better information regarding the dispersion. The concept thus discussed so far is summarized in Figure 2.13. In conclusion it can be said that the probability system is wholly abstract and axiomatic. Consequently, every fully defined probability problem has a unique solution.

2.7 DISCRETE PROBABILITY DISTRIBUTIONS

In probability theory, there are several rules that define the functional relationships between the possible values of random variable *X* and their probabilities, *P(X)*. As they are purely theoretical, i.e. they do not exist in reality, they are called *theoretical probability distributions*. Instead of analyzing the ways in which these rules have been derived, the analysis in this chapter concentrates on their properties. It is necessary to emphasize that all theoretical distributions represent the family of distributions defined by a common rule through unspecified constants known as *parameters of distribution*. The particular member of the family is defined by fixing numerical values for the parameters, which define the distribution. The probability distributions most frequently used in reliability and Six Sigma are examined in this chapter.

Among the family of theoretical probability distributions that are related to discrete random variables, the Binomial distribution and the Poisson distribution are relevant to the objectives set by this book. A brief description of these two distributions is given below.

2.7.1 Bernoulli trials

The simple probability distribution is one with only two event classes. For example, a car is tested and one of two events, pass or fail, must occur, each with some probability. The type of experiment consisting of series of independent trials, each of which can eventuate in only one of two outcomes are known as *Bernoulli Trials*, and the two event classes and their associated probabilities a *Bernoulli Process*. In general, one of the two events is called a "success" and the other a "failure" or "nonsuccess". These names serve only to tell the events apart, and are not meant to bear any connotation of "goodness" of the event. The symbol p, stands for the probability of a success, q for the probability of failure $(p + q = 1)$. If 5 independent trials are made $(n = 5)$, then $2^5 = 32$ different sequences of possible outcomes (in general 2^n) would be observed.

The probability of given sequences depends upon p and q, the probability of the two events. Fortunately, since trials are independent, it is possible to compute the probability of any sequence.

If all possible sequences and their probabilities are written down the following fact emerges: *The probability of any given sequences of n independent Bernoulli Trials depends only on the number of successes and p*. This is regardless of the order in which successes and failure occur in a sequence. The corresponding probability is:

$$p^r q^{n-r}$$

Where r is the number of successes, and $n-r$ is the number of failures. Suppose that in a sequence of 10 trials, exactly 4 successes occur. Then the probability of that particular sequence is $p^4 q^6$. If $p = \frac{2}{3}$, then the probability can be worked out from:

$$\left(\frac{2}{3}\right)^4\left(\frac{1}{3}\right)^6$$

The same procedure is followed for any r successes out of n trials and for any p. Generalizing this idea for any r, n, and p, we have the following principle:

In sampling from the Bernoulli Process with the probability of a success equal to p, the probability of observing exactly r successes in n independent trials is:

$$P(r\ successes|n,p) = \binom{n}{r} p^r q^{n-r} = \frac{n!}{r!(n-r)!} p^r q^{n-r} \tag{2.27}$$

2.7.2 Binomial distribution

The theoretical probability distribution, which pairs the number of successes in n trials with its probability, is called the binominal distribution.

This probability distribution is related to experiments, which consist of a series of independent trials, each of which can result in only one of two outcomes: success or failure. By convention the symbol p stands for the probability of a success, q for the probability of failure $(p+q=1)$.

The number of successes, x in n trials is a discrete random variable which can take on only the whole values from 0 through n. The PMF of the Binomial distribution is given by:

$$PMF(x) = P(X=x) = \binom{n}{x} p^x q^{n-x}, \qquad 0<x<n \tag{2.28}$$

where:

$$\binom{n}{x} p^x q^{n-x} = \frac{n!}{x!(n-x)!} p^x q^{n-x} \tag{2.29}$$

The binomial distribution expressed in cumulative form, representing the probability that X falls at or below a certain value 'a' is defined by the following equation:

$$P(X \leq a) = \sum_{i=o}^{a} P(X = x_i) = \sum_{i=0}^{a} \binom{n}{i} p^i q^{n-i} \tag{2.30}$$

$$E(X) = np \tag{2.31}$$

Similarly, because of the independence of trials, the variance of the binomial distribution is the sum of the variances of the individual trials, or $p(1-p)$ summed n times:

$$V(X) = np(1-p) = npq \tag{2.32}$$

Consequently, the standard deviation is equal to:

$$Sd(X) = \sqrt{npq} \tag{2.33}$$

As an illustration of the binomial distribution, the PMF and CDF are shown in Figure 2.14 with parameters $n = 10$ and $p = 0.3$.

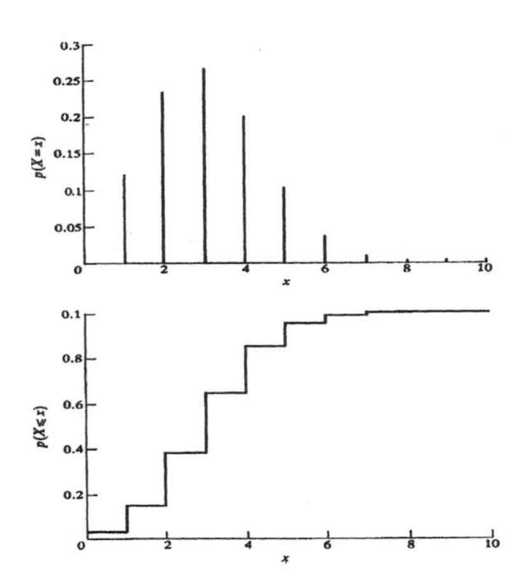

Figure 2-14. PMF and CDF For Binomial Distribution, n = 10, p = 0.3

Although the mathematical rule for the binomial distribution is the same regardless of the particular values which parameters n and p take, the shape of the probability mass function and the cumulative distribution function will depend upon them. The *PMF* of the binomial distribution is symmetric if $p = 0.5$, positively skewed if $p < 0.5$, and negatively skewed if $p > 0.5$.

2.7.3 Poisson distribution

The theoretical probability distribution which pairs the number of occurrences of an event in a given time period with its probability is called the Poisson distribution. There are experiments where it is not possible to observe a finite sequence of trials. Instead, observations take place over a continuum, such as time. For example, if the number of ions moving between plates in a battery in a given period of time is observed, say for one minute, it is difficult to think of this situation in terms of finite trials. If the number of binomial trials *n,* is made larger and larger and *p* smaller and smaller in such a way that *np* remains constant, then the probability distribution of the number of occurrences of the random variable approaches the Poisson distribution.

The probability mass function in the case of the Poisson distribution for random variable *X* can be expressed as follows:

$$P(X = x|\lambda) = \frac{e^{-\lambda}\lambda^{x}}{x!} \qquad \text{where } x = 0, 1, 2, \ldots \tag{2.34}$$

λ is the *intensity of the process* and represents the expected number of occurrences in a time period of length *t*. Figure 2.15 shows the *PMF* of the Poisson distribution with $\lambda = 5$

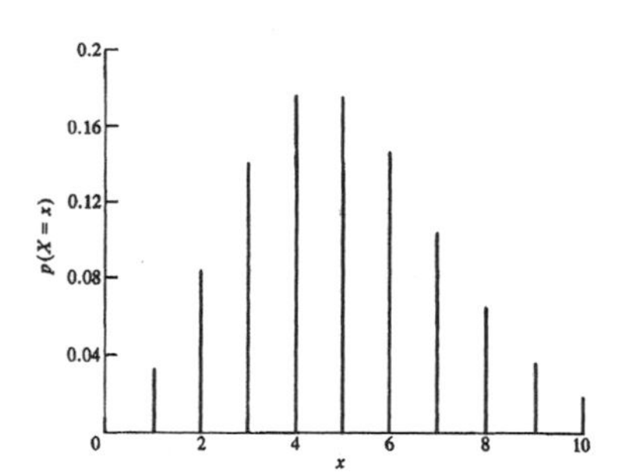

Figure 2-15. PMF of the Poisson distribution for $\lambda = 5$

The Cumulative Distribution Function for the Poisson distribution is given by:

$$F(x) = P(X \leq x) = \sum_{i=o}^{x} \frac{e^{-\lambda} \lambda^{i}}{i!} \tag{2.35}$$

The CDF of the Poisson distribution with $\lambda = 5$ is presented in Figure 2.16. Expected value of the distribution is given by

$$E(X) = \sum_{x=0} xP(X = x) = \sum_{x=0} x \frac{e^{-\lambda} \lambda^{x}}{x!}$$

Applying some simple mathematical transformations it can be proved that:

$$E(X) = \lambda \tag{2.36}$$

That is the expected number of occurrences in a period of time t is equal to λ. The variance of the Poisson distribution is equal to the mean:

$$V(X) = \lambda \tag{2.37}$$

Thus, the Poisson distribution is a single parameter distribution because it is completely defined by the parameter λ. In general, the Poisson distribution is positively skewed, although it is nearly symmetrical as λ becomes larger.

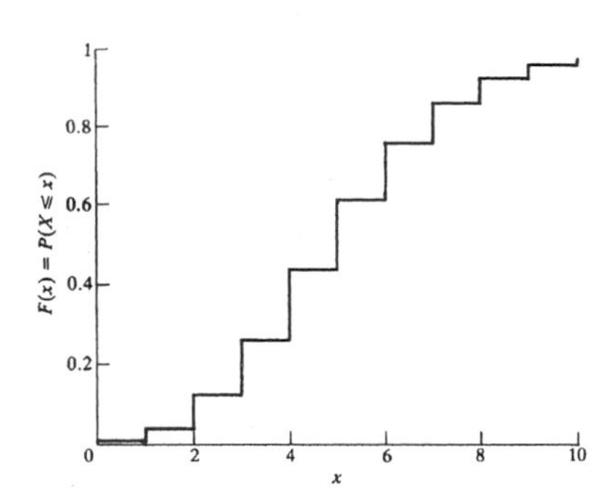

Figure 2-16. CDF of the Poisson distribution λ = 5

The Poisson distribution can be derived as a limiting form of the binomial if the following three assumptions were simultaneously satisfied:

1. n becomes large (that is, $n \to \infty$).
2. p becomes small (that is, $p \to 0$).

3. np remains constant.

Under these conditions, the binomial distribution with the parameters n and p, can be approximated by the Poisson distribution with parameter $\lambda = np$. This means that the Poisson distribution provides a good approximation to the binomial distribution if p is very small and n is large. Since p and q can be interchanged by simply interchanging the definitions of success and failure, the Poisson distribution is also a good approximation when p is close to one and n is large.

As an example of the use of the Poisson distribution as an approximation to the binomial distribution, the case in which n = 10 and p = 0.10 may be considered. The Poisson parameter for the approximation is then $\lambda = np = 10 \times 0.10 = 1$. The binomial distribution and the Poisson approximation are shown in Table 2.1.

The two distributions agree reasonably well. If more precision is desired, a possible rule of thumb is that the Poisson is a good approximation to the binomial if $n / p > 500$ (this should give accuracy to at least two decimal places).

Table 2-1. Poisson distribution as an approximation to the binomial distribution

x	Binomial $P(X = x \| n = 10, p = 0.1)$	Poisson $P(X = x \| \lambda = 1)$
0	0.598737	0.606531
1	0.315125	0.303265
2	0.074635	0.075816
3	0.010475	0.012636
4	0.000965	0.001580
5	0.000061	0.000158

2.8 CONTINUOUS PROBABILITY DISTRIBUTIONS

It is necessary to emphasize that all theoretical distributions represent the family of distributions defined by a common rule through unspecified constants known as *parameters of distribution*. The particular member of the family is defined by fixing numerical values for the parameters, which define the distribution. The probability distributions most frequently used in reliability and Six Sigma are examined in this chapter. Each of the above mentioned rules define a family of distribution functions. Each member of the family is defined with a few parameters, which in their own way control the distribution. Parameters of a distribution can be classified in the following three categories (note that not all distributions will have all the three parameters, many distributions may have either one or two parameters):

1. *Scale parameter*, which controls the range of the distribution on the horizontal scale.
2. *Shape parameter*, which controls the shape of the distribution curves.
3. *Source parameter or Location parameter*, which defines the origin or the minimum value which random variable, can have. Location parameter also refers to the point on horizontal axis where the distribution is located.

Thus, individual members of a specific family of the probability distribution are defined by fixing numerical values for the above parameters.

2.8.1 Exponential distribution

Exponential distribution is fully defined by a single one parameter that governs the scale of the distribution. The probability density function of the exponential distribution is given by:

$$f(x) = \lambda \exp(-\lambda x), x > 0 \tag{2.38}$$

In Figure 2.17 several graphs are shown of exponential density functions with different values of λ. Notice that the exponential distribution is positively skewed, with the mode occurring at the smallest possible value, zero.

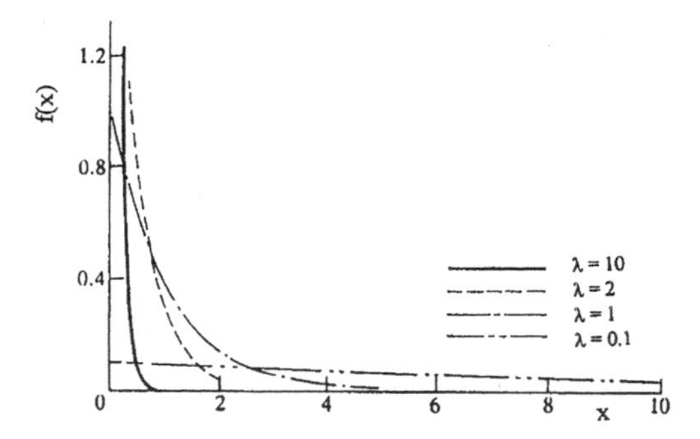

Figure 2-17. Probability density function of exponential distribution for different values of λ

The cumulative distribution of exponential distribution is given by:

$$F(x) = P(X < x) = 1 - \exp(-(\lambda x)) \tag{2.39}$$

It can be shown that the mean and variance of the exponential distribution are:

$$E(X) = 1/\lambda \tag{2.40}$$

$$V(X) = (1/\lambda)^2 \tag{2.41}$$

The standard deviation in the case of the exponential distribution rule has a numerical value identical to the mean, $SD(X) = E(X) = 1/\lambda$.

2.8.1.1 Memory less property

One of the unique properties of exponential distribution is that it is the only continuous distribution that has *memory less* property. Suppose that the random variable X measures the duration of time until the occurrence of failure of an item and that it is known that X has an exponential distribution with parameter λ. Suppose the present age of the item is t, that is $X > t$. Assume that we are interested in finding the probability that this item will not fail for another s units of time. This can be expressed using the conditional probability as:

$$P\{X > s+t | X > t\}$$

Using conditional probability of events, the above probability can be written as:

$$P\{X > s+t | X > t\} = \frac{P\{X > s+t \cap X > t\}}{P\{X > t\}} = \frac{P\{X > s+t\}}{P\{X > t\}} \tag{2.42}$$

However we know that for exponential distribution

$$P[X > s+t] = \exp(-\lambda(s+t)) \text{ and } P[X > t] = \exp(-\lambda t)$$

Substituting these expressions in equation (2.42), we get

$$P[X > s+t | X > t] = P[X > s] = \exp(-\lambda s)$$

That is, the conditional probability depends only on the remaining duration and is independent of the current age of the item. *This property is exploited to a great extent in reliability theory.*

2.8.2 Normal distribution

This is the most frequently used and most extensively covered theoretical distribution in the literature. The foundations of Six Sigma are based on the normal distribution. The Normal Distribution is continuous for all values of X between $-\infty$ and $+\infty$. It has a characteristic symmetrical shape, which means that the mean, the median and the mode have the same numerical value. The mathematical expression for its probability density function is as follows:

$$f(x) = \frac{1}{\sigma\sqrt{2\pi}} \exp\left(-\frac{1}{2}\left(\frac{x-\mu}{\sigma}\right)^2\right) \tag{2.43}$$

Where μ is the location parameter (as it locates the distribution on the horizontal axis), σ is the scale parameter (as it controls the range of the distribution). In a normal distribution, μ and σ also represents the mean and the standard deviation. The influence of the parameter μ on the location of the distribution on the horizontal axis is shown in Figure 2.18, where the values for parameter σ are constant.

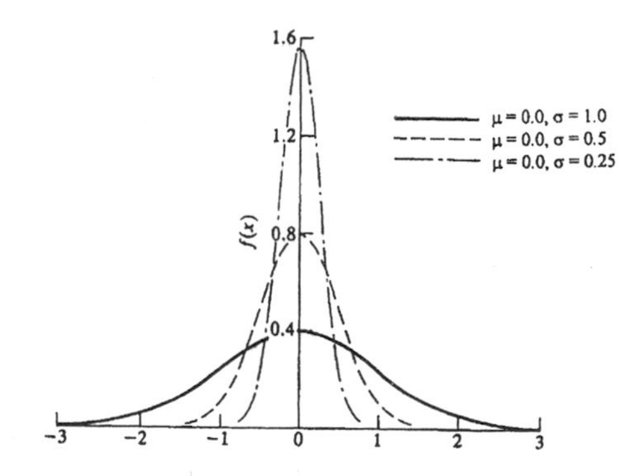

Figure 2-18. Probability density of normal distribution for different σ values

As the deviation of x from the location parameter μ is entered as a squared quantity, *two* different x values, showing the same absolute deviation from μ, will have the same probability density according to this rule. This dictates the symmetry of the normal distribution. Parameter μ can be any finite number, while σ can be any *positive* finite number. The cumulative distribution function for the normal distribution is:

$$F(a) = P(X \leq a) = \int_{-\infty}^{a} f(x)dx$$

Where *f(x)* is the probability density function. Taking into account equation (2.43) this becomes:

$$F(a) = \int_{-\infty}^{a} \frac{1}{\sigma\sqrt{2\pi}} \exp\left(-\frac{1}{2}\left(\frac{x-\mu}{\sigma}\right)^2\right)dx \tag{2.44}$$

In Figure 2.19 several cumulative distribution functions are given of the Normal Distribution, corresponding to different values of μ *and* σ. As the integral in equation (2.44) cannot be evaluated in a closed form, statisticians have constructed the table of probabilities, which comply with the normal rule for the standardized random variable, Z. Z is a theoretical random variable with parameters $\mu = 0$ and $\sigma = 1$. The relationship between standardized random variable Z and random variable X is established by the following expression:

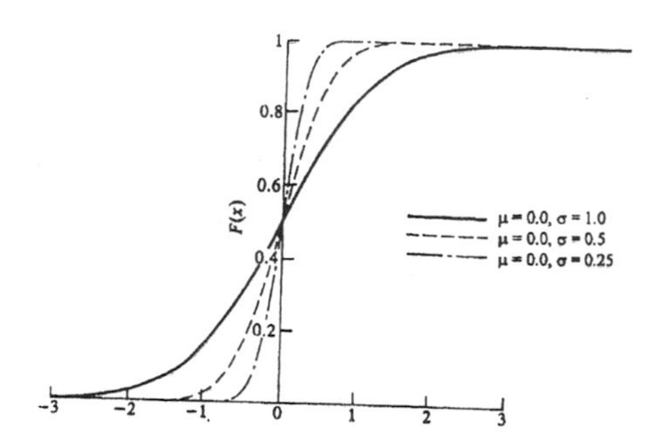

Figure 2-19. Cumulative distribution of normal distribution for different values of μ and σ

$$z = \frac{x - \mu}{\sigma} \tag{2.45}$$

Making use of the above expression the equation (2.43) becomes simpler:

$$f(z) = \frac{1}{\sigma\sqrt{2\pi}} e^{-\frac{1}{2}z^2} \tag{2.46}$$

The standardized form of the distribution makes it possible to use only one table for the determination of *PDF* for any normal distribution, regardless of its particular parameters (see Table in appendix). The relationship between *f(x)* and *f(z)* is :

$$f(x) = \frac{f(z)}{\sigma} \tag{2.47}$$

By substituting $\frac{x - \mu}{\sigma}$ in place of *z*, equation. (2.44) becomes:

$$F(a) = \int_{-\infty}^{a} \frac{1}{\sigma\sqrt{2\pi}} \exp\left(-\frac{1}{2} z^2 \right) dz = \Phi\left(\frac{a - \mu}{\sigma} \right) \tag{2.48}$$

Where Φ is the standard normal distribution function and is given by:

$$\Phi(z) = \int_{-\infty}^{x} \frac{1}{\sqrt{2\pi}} \exp\left(-\frac{1}{2} z^2 \right) dx \tag{2.49}$$

The corresponding standard normal probability density function is:

$$f(z) = \frac{1}{\sqrt{2\pi}} \exp\left(-\frac{z^2}{2} \right) \tag{2.50}$$

Most tables of the normal distribution give the cumulative probabilities for various *standardized* values. That is, for a given *z* value the table provides the cumulative probability up to and including that standardized value in a normal distribution. In *Microsoft EXCEL®*, the cumulative distribution function and density function of normal distribution with mean μ and standard deviation σ can be found using the following function.

$F(x) = NORMDIST(x, \mu, \sigma, TRUE)$, *and* $f(x) = NORMDIST(x, \mu, \sigma, FALSE)$

If F(z) value is known, then to find the value of z using *Microsoft EXCEL* the following function may be used.

z = NORMINV(F(z), 0, 1)

Note that the value of z is nothing but the Sigma level quality as measured in Six Sigma measurement system (see chapter 3 for more details). The expected value of the normal random variable is equal to the location parameter, μ. That is:

$$E(X) = \mu \tag{2.51}$$

Whereas the variance is

$$V(X) = \sigma^2 \tag{2.52}$$

Since normal distribution is symmetrical about its mean, the area between μ - kσ, μ + kσ (k is any real number) takes a unique value, which is shown in Figure 2.20

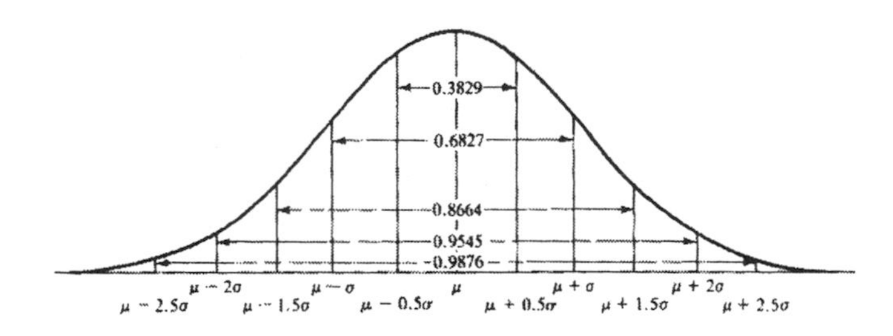

Figure 2-20. The areas under a normal distribution between μ- kσ and μ+kσ

Table 2.2 gives the area between μ - kσ, μ + kσ for k = 1, 2, …, 6.

Table 2-2. Area between μ - kσ, μ + kσ under a normal distribution

Range	Area (cumulative probability)
μ - 1 σ and μ + 1 σ	0.68268948
μ - 2 σ and μ + 2σ	0.954499876
μ - 3σ and μ + 3σ	0.997300066
μ - 4σ and μ + 4σ	0.999936628
– ∞ and μ + 4.5 σ	0.999996599
μ - 5σ and μ + 5σ	0.999999426
μ - 6σ and μ + 6σ	0.999999998

The area between – ∞ and μ + 4.5 σ gives the defect rate of 3.4 out of one million which is the target for Six Sigma quality.

2.8.2.1 Central limit theorem

Suppose $X_1, X_2, \ldots X_n$ are mutually independent observations of a random variable X having a well-defined mean μ_x and standard deviation σ_x. Let

$$Z_n = \frac{\bar{X} - \mu_x}{\sigma_x / n} \tag{2.53}$$

Where,

$$\bar{X} = \frac{1}{n} \sum_{i=1}^{n} X_i \tag{2.54}$$

and $F_{z_n}(z)$ be the cumulative distribution function of the random variable Z_n. Then for all z, $-\infty < z < \infty$,

$$\lim_{n \to \infty} F_{Z_n}(z) = F_Z(z) \tag{2.55}$$

Where $F_Z(z)$ is the cumulative distribution of standard normal distribution *N(0,1)*. The X values have to be from the same distribution but the remarkable feature is that this distribution does not have to be normal, it can be uniform, exponential, beta, gamma, Weibull or even an unknown one.

2.8.3 Lognormal distribution

The lognormal probability distribution, can in some respect, be considered as a special case of the normal distribution because of the derivation of its probability function. If a random variable $Y = \ln X$ is normally distributed then, the random variable X follows the lognormal distribution. Thus, the probability density function of a log-normal random variable X is defined as:

$$f_X(x) = \frac{1}{x\sigma_l\sqrt{2\pi}}\exp\left(-\frac{1}{2}\left(\frac{\ln x - \mu_l}{\sigma_l}\right)^2\right) \quad \geq 0 \qquad (2.56)$$

The parameter μ_l is called the *scale parameter* (see Figure 2.21) and parameter σ_l is called the *shape parameter*. The relationship between parameters μ (location parameter of the normal distribution) and μ_l is defined by the following expression:

$$\mu = \exp\left(\mu_l + \frac{1}{2}\sigma_l^2\right) \qquad (2.57)$$

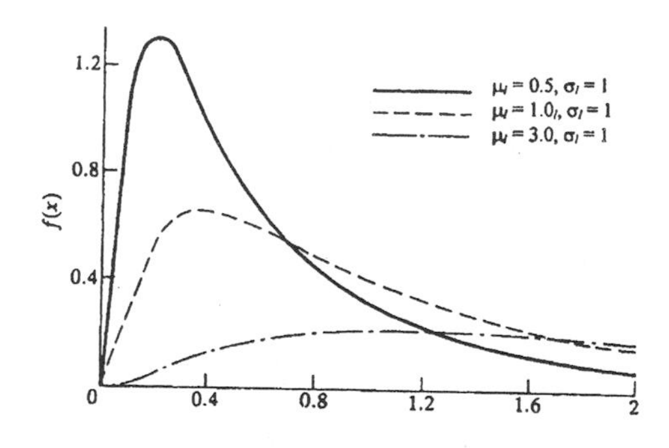

Figure 2-21. Probability density of log-normal distribution

The cumulative distribution function for the lognormal distribution is defined by the following expression:

$$F_X(x) = P(X \leq x) = \int_0^x \frac{1}{x\sigma_l\sqrt{2\pi}} \exp\left(-\frac{1}{2}\left(\frac{\ln x - \mu_l}{\sigma_l}\right)^2\right) dx \quad (2.58)$$

As the integral cannot be evaluated in closed form, the same procedure is applied as in the case of normal distribution to calculate the cumulative distribution of a log-normal distribution. Thus, making use of the standardized random variable, equation (2.58) transforms into:

$$F_X(x) = P(X \leq x) = \Phi\left(\frac{\ln x - \mu_l}{\sigma_l}\right) \quad (2.59)$$

The measures of central tendency in the case of lognormal distributions are defined by the:

(a) Location parameter (Mean)

$$M = E(X) = \exp\left(\mu_l + \frac{1}{2}\sigma_l^2\right) \quad (2.60)$$

(b) Deviation parameter (the variance)

$$V(X) = \exp\left(2\mu_l + \sigma_l\right)^2\left[\exp(\sigma_l^2 - 1)\right] \quad (2.61)$$

2.8.4 Weibull distribution

This distribution originated from the experimentally observed variations in the yield strength of Bofors steel, the size distribution of fly ash, fiber strength of Indian cotton, and the fatigue life of a *St*-37 steel by the Swedish engineer W.Weibull. As the Weibull distribution has no characteristic shape, such as the normal distribution, it has a very important role in the statistical analysis of experimental data. The shape of this distribution is governed by its parameter. The rule for the probability density function of the Weibull distribution is:

$$f(x)=\frac{\beta}{\eta}\left(\frac{x-\gamma}{\eta}\right)^{\beta-1}\exp\left[-\left(\frac{x-\gamma}{\eta}\right)^{\beta}\right],\qquad \eta,\beta,\gamma\geq 0\qquad x\geq\gamma \tag{2.62}$$

where η, β, $\gamma > 0$. As the location parameter γ is often set equal to zero, in such cases:

$$f(x)=\frac{\beta}{\eta}\left(\frac{x}{\eta}\right)^{\beta-1}\exp\left[-\left(\frac{x}{\eta}\right)^{\beta}\right]$$

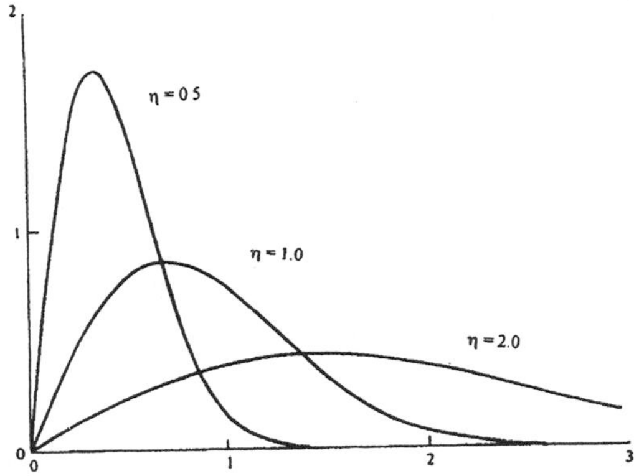

Figure 2-22. Probability density of Weibull distribution with $\beta = 2.0$, $\gamma = 0$, $\eta = 0.5, 1, 2$

For different parameter values of β, the Weibull distribution takes different shapes. For example, when $\beta = 3.4$ the Weibull approximates to the normal distribution; when $\beta=1$, it is identical to the exponential distribution. Figure 2.22 shows the Weibull probability density function for selected parameter values. The cumulative distribution functions for the Weibull distribution is:

$$F(x)=1-\exp\left[-\left(\frac{x-\gamma}{\eta}\right)^{B}\right] \tag{2.63}$$

For $\gamma = 0$, the cumulative distribution is given by

$$F(x) = 1 - \exp\left[-\left(\frac{x}{\eta}\right)^{\beta}\right] \tag{2.64}$$

The expected value of the Weibull distribution is given by:

$$E(X) = \gamma + \eta \times \Gamma\left(\frac{1}{\beta} + 1\right) \tag{2.65}$$

where Γ is the gamma function, defined as

$$\Gamma(n) = \int_0^{\infty} e^{-x} \times x^{n-1} dx$$

When n is integer then $\Gamma(n) = (n-1)!$. For other values, one has to solve the above integral to the value. Values for this can be found in Gamma function table given in the appendix. In *Microsoft EXCEL,* Gamma function, $\Gamma(x)$ can be found using the function, *EXP[GAMMALN(x)].* The variance of the Weibull distribution is given by:

$$V(X) = (\eta)^2\left[\Gamma\left(1 + \frac{2}{\beta}\right) - \Gamma^2\left(1 + \frac{1}{\beta}\right)\right] \tag{2.66}$$

2.9 STOCHASTIC PROCESSES

Stochastic process (also known as random process) is a collection of random variables $\{X(t), t \in T\}$, where *T* is the set of numbers that indexes the random variables X(t). In reliability, it is often appropriate to interpret t as time and T as the range of time being considered. The set of possible values the stochastic process X(t) can assume is called state. The set of possible states constitutes the state-space, denoted by E. The state-space can be continuous or discrete. For example consider a system with two items connected in parallel. Assume that the time-to-failure distributions of the two parallel items are given by two sequences of random variables X_i and Y_i (i = 1, 2 ...). Here the subscript i represents the time to i^{th} failure of the items. If the sequence of random variable Z_i represents the ith repair time, then the process $\{ X(t), t \geq 0 \}$ by definition forms a stochastic process. At any time t, it is possible that two, one or none of these two items will be maintaining the required function. Thus, the set {0, 1, 2} forms the state-space of the system.

Analyzing a system using stochastic processes will involve the following fundamental steps.

1. Identify the time domain T for the system. The time domain T can be discrete or continuous.
2. Identify the state space of the system. The state space can be either discrete or continuous.

Once the process is defined using the family of random variables $\{X(t), t \in T\}$, state space (E) and the parameter set (T), the next step will be to identify the properties of the process that can be used to classify the process and also to analyze the process to extract information. As far as reliability is concerned, processes with a continuous time parameter and discrete state space are important. In this chapter, we discuss the following stochastic processes.

1. Markov processes
2. Non-homogeneous Poisson Processes
3. Renewal processes

Readers who are interested to know more on applications of stochastic process are advised to refer to Birolini *(1997)*.

2.10 MARKOV PROCESSES

A stochastic process is said to be a Markov process if the future evolution of the process depends only on the current time and state. That is, the future state of a system is conditionally independent of the past, given that the present state and age of the system is known. Thus, to predict the future state one need to know only the present state and age of the system. Mathematically, a stochastic process $\{X(t); t \in T\}$ with state-space E is called a Markov process if it satisfies the condition:

$$\begin{aligned} P[X(t_n+h)=j \mid X(t_n)=i_n, X(t_{n-1})=i_{n-1},\ldots,X(t_0)=i_0] \\ = P[X(t_n+h)=j \mid X(t_n)=i_n] \end{aligned} \tag{2.67}$$

for all $(j, i_n, i_{n-1}, \ldots, i_0) \in E$. The above property is called *Markov property*. A Markov process with discrete state space is called *Markov Chain*. A Markov process with continuous time and discrete state space is called *continuous time Markov chain (CTMC)*. The conditional probability defined in equation (2.67) is referred as the *transition probability* of Markov process and is defined using the notation $P_{ij}(t_n + h)$

$$P_{ij}(t_n + h) = P[X(t_n + h) = j \mid X(t_n) = i_n] \tag{2.68}$$

A Markov process is called *time-homogeneous* or *stationary* if the transition probabilities are independent of time t. For a stationary Markov process,

$$P_{ij}(t_n + h) = P_{ij}(t_0 + h) = P_{ij}(h) \tag{2.69}$$

Thus, the transition from state i to state j in a stationary Markov chain depends only on the duration h. The transition probabilities $P_{ij}(t + h)$ satisfy the following *Chapman-Kolmogrov* equations

$$P_{ij}(t + h) = \sum_{k \in E} P_{ik}(h) P_{kj}(t) \tag{2.70}$$

In all the models discussed in this Chapter we assume that the Markov process is stationary. It is convenient to use a matrix to represent various state transition probabilities of a Markov process. For example, if a system has n states, we define a matrix **P**, such that

$$\mathbf{P} = [P_{ij}(h)] = \begin{bmatrix} P_{11} & P_{12} & \cdots & P_{1n} \\ P_{21} & P_{22} & \cdots & P_{2n} \\ . & . & . & . \\ . & . & . & . \\ P_{n1} & P_{n2} & \cdots & P_{nn} \end{bmatrix}$$

The matrix **P** is called *Transition Probability Matrix* (TPM) or *Stochastic Matrix*.

Let $\{S_j, j \in E\}$ represent the time spent at state *j* (sojourn time at state *j*). The probability that the process will spend more than *t* hours at state *j* is, $P[S_j > t]$. Assume that the process has already spent *h* hours in state *j*, the probability that it will spend additional *t* hours in state *j* is given by:

$$P[S_j > t + h \mid S_j > h] \tag{2.71}$$

Since past is irrelevant in Markov process, the above expression can be written as:

$$P[S_j > t + h \mid S_j > h] = P[S_j > t] \quad (2.72)$$

The only continuous distribution that satisfies the above relation is exponential distribution. The above property of exponential distribution is called *memory-less property.* Thus, in a Markov process, the time spent in any state follows exponential distribution. Thus,

$$P[S_j > t] = \exp(-\nu_j t) \quad (2.73)$$

where the parameter ν_i depends on state *i.* This is a very important result and limitation of Markov processes. This implies that the Markov process can be applied in reliability theory only when the time-to-failure follow exponential distribution.

Transition Rates between the States of a Markov Process

Since the time spent at any state *j* of a Markov process follows exponential distribution, the probability that the process remains in state *j* during a small interval δt is given by:

$$\begin{aligned} P[S_j > \delta t] &= \exp(-\nu_j \delta t) \\ &= 1 - \frac{\nu_j \delta t}{1!} + \frac{(\nu_j \delta t)^2}{2!} - \ldots \\ &= 1 - \nu_j \delta t + O(\delta t) \end{aligned}$$

where $O(\delta t)$ represents the terms which are negligible as δ approaches zero. That is,

$$\lim_{\delta t \to 0} \frac{O(\delta t)}{\delta t} = 0$$

Thus, for a small duration of δt, $P_{jj}(\delta t)$, probability that the process will remain in state *j* for small duration δt is given by:

$$P_{jj}(\delta t) = 1 - \nu_j \delta t + O(\delta t)$$

Probability that the system will leave state *j* is given by

$$1 - P_{jj}(\delta t) = v_j \delta t + O(\delta t)$$

v_j is the rate at which the process $\{X(t),\ t \in T\}$ leaves the state *j*. Rearranging the above equation we have,

$$P_{jj}(\delta t) - 1 = -v_j \delta t + O(\delta t)$$

Substituting $\lambda_{jj} = -v_j$ in the above equation, we get

$$P_{jj}(\delta t) - 1 = \lambda_{ii} \delta t + O(\delta t)$$

It is easy to verify that

$$\lim_{\delta t \to 0} \frac{P_{jj} - 1}{\delta t} = \lambda_{jj} \tag{2.74}$$

The transition probability $P_{ij}(\delta t)$, that is the process will enter state *j* (with probability r_{ij}) after leaving state *i* during a small duration δt is given by:

$$\begin{aligned} P_{ij}(\delta t) &= [1 - P_{ii}(\delta t)] \times r_{ij} = [v_i \delta t + O(\delta t)] \times r_{ij} \\ &= \lambda_{ij} \delta t + O(\delta t) \end{aligned} \tag{2.75}$$

where λ_{ij} is the rate at which the process enters the state *j* from the state *i*.

Let $P_j(t) = P[X(t) = j]$, that is $P_j(t)$ denotes that the process is in state *j* at time t. Now for any δt, we have

$$\begin{aligned} P_j(t + \delta t) &= P[X(t + \delta t) = j] \\ &= \sum_{i \in E} P[X(t + \delta t) = j \mid X(t) = i] P[X(t) = i] \end{aligned}$$

The above expression can be written as

$$P_j(t + \delta t) = \sum_{i \in E} P_{ij}(\delta t) P_{i(t)} \tag{2.76}$$

The above equation (2.76), upon few mathematical manipulation will give a system of differential equations which can be solved to find $P_i(t)$.

From equation (2.76)

$$P_j(t+\delta t) - P_j(t) = \sum_{\substack{i \in E \\ i \neq j}} P_{ij}(\delta t) P_i(t) + P_j(t)[P_{jj}(\delta t) - 1] \tag{2.77}$$

For $\delta t \to 0$, and using equation (2.74) and (2.75), equation (2.77) can be written as:

$$\frac{d}{dt} P_j(t) = \sum_{i \in E} \lambda_{ij} P_i(t) = \sum_{\substack{i \in E \\ i \neq J}} \lambda_{ij} P_i(t) - v_j P_j(t) \tag{2.78}$$

Also

$$\sum_{j \in E} P_j(t) = 1 \tag{2.79}$$

Equation (2.78) is called *Kolmogrov backward equations*, which along with equation (2.79) has a unique solution. Thus, various state probabilities of the process can be obtained by solving the system of differential equations of the form:

$$\frac{d}{dt} P(t) = \Delta P(t) \tag{2.80}$$

where P(t) is a time-dependent N dimensional probability vector and Δ is a square matrix where the element (i,j) represents the rate at which the process enters the state *j* from the state *i*.

2.11 NON-HOMOGENEOUS POISSON PROCESS

A counting process $\{N(t),\ t \geq 0\}$ is said to be a *non-homogeneous Poisson process* with intensity function $\lambda(t),\ t \geq 0$, if:

1. $N(0) = 0$.

2. $N(t)$ has independent increments.
3. The number of events in any interval t and $t+s$ has a Poisson distribution with mean $[S(t+s) - S(t)]$, that is

$$P[N(t+s)-N(t)=n]=\frac{[S(t+s)-S(t)]^n \exp\{-(S(t+s)-S(t))\}}{n!} \tag{2.81}$$

Where

$$S(t)=\int_0^t \lambda(x)dx \tag{2.82}$$

$S(t)$ is the expected number of events in (0,t). Also, $N(t+s) - N(t)$ is Poisson distributed with mean $S(t+s) - S(t)$.

2.12 RENEWAL PROCESS

Renewal theory was originally used to analyze the replacement of equipment upon failure, to find the distribution of number of replacement and mean number of replacements. Let $\{X_n;\ n = 1, 2, \ldots\}$ be a sequence of non-negative independent random variables with common distribution F. Let X_n be the time between $(n-1)^{st}$ and n^{th} event. Let:

$$S_0=0,\ S_n=\sum_{i=1}^{n} X_i \tag{2.83}$$

Thus S_n is the time to n^{th} event or *epoch* at which the nth renewal occurs. Let N(t) be the number of renewals by time t.

$$N(t)=Max\{n; \leftrightarrow \quad S_n \le t\} \tag{2.84}$$

Let X_1, X_{2},, ... are independent and identically distributed random variables with distribution $F(t)$. Then $P\{S_n \le t\}$ is given by:

$$P\{S_n(t)\le t\}=F^n(t) \tag{2.85}$$

where $F^n(t)$ is the n-fold convolution of $F(t)$. That is,

$$F^n(t) = \int_0^t F^{n-1}(x)dF(x) \tag{2.86}$$

We use the convention that $F^0(t) = 1$ for $t > 0$. $F^n(t)$ represents the probability that the nth renewal occurs by time t. The distribution of N(t) can be derived using the following arguments.

Distribution of N(t)

The counting process, N(t), is called a renewal process. From the definition of N(t) and S_n, we have

$$\{N(t) = n\} \Leftrightarrow \{S_n \le t, S_{n+1} > t\} \tag{2.87}$$

$$\begin{aligned} P[N(t) = n] &= P\{N(t) < n+1\} - P\{N(t) < n\} \\ &= P\{S_{n+1} > t\} - P\{S_n > t\} \\ &= 1 - F^{n+1}(t) - [1 - F^n(t)] \end{aligned} \tag{2.88}$$

Thus the probability that the number of renewal by time t is equal to n, is given by:

$$P\{N(t) = n\} = F^n(t) - F^{n+1}(t) \tag{2.89}$$

It is difficult to evaluate the above function analytically for many theoretical distributions, however it can be solved using well-known numerical methods.

2.12.1 Renewal Function

The expected number of renewals during specified duration t is given by:

$$E[N(t)] = M(t) = \sum_{i=1}^{\infty} i \times [F^i(t) - F^{i+1}(t)] \tag{2.90}$$

The above equation can be simplified, and the expected number of renewals (expected number of demands) is given by:

$$M(t) = \sum_{i=1}^{\infty} F^i(t) \tag{2.91}$$

The above equation is called *renewal function, M(t),* and it gives the number of renewals during (0, t]. Taking the derivative of renewal function we get:

$$m(t) = \frac{d}{dt} M(t) = \sum_{n=1}^{\infty} f^n(t) \tag{2.92}$$

Where $f^n(t)$ is the derivative of $F^n(t)$. $m(t)\delta t$ is the probability that a renewal occurs during (t, t+δt). *m(t)* is called the *renewal density* or *renewal rate.*

Calculating $F^n(t)$, P[N(t) = n], M(t) and m(t)

Exponential Distribution

$$F(t) = 1 - \exp(-\lambda t)$$

When the time to failure distribution is exponential, the renewal process constitutes a Poisson process. Thus, Poisson process is also a special case of renewal process where time to failure is exponential.

$$F^n(t) = 1 - \sum_{i=0}^{n-1} \frac{\exp(-\lambda t) \times (\lambda t)^i}{i!} \tag{2.93}$$

$$P[N(t) = n] = \frac{\exp(-\lambda t) \times (\lambda t)^n}{n!} \tag{2.94}$$

$$M(t) = \lambda t \tag{2.95}$$

$$m(t) = \lambda \tag{2.96}$$

Normal Distribution

By assuming $\sigma << \mu$, we have

$F^n(t) = \Phi(\frac{t - n \times \mu}{\sigma \times \sqrt{n}})$, where $\Phi(t)$ is the standard normal distribution.

The distribution of N(t) is given by:

$$P[N(t) = n] = \sum_{n=1}^{\infty}[\Phi(\frac{t - n \times \mu}{\sigma \times \sqrt{n}}) - \Phi(\frac{t - (n+1) \times \mu}{\sigma \times \sqrt{n+1}})]$$

$$M(t) = \sum_{n=1}^{\infty} \Phi(\frac{t - n \times \mu}{\sigma \times \sqrt{n}}) \tag{2.97}$$

For distributions like Weibull, one has to use numerical approximation to find the renewal function.

2.12.2 Elementary Renewal Theorem

For a distribution function F(t) with F(0) = 0 and finite mean, and if f(x) exists then the following equation is valid

$$\lim_{t \to \infty} \frac{M(t)}{t} = \frac{1}{MTTF} \tag{2.98}$$

The above result is called the *Elementary Renewal Theorem.* This implies that in the steady state, the expected number of failures is given by the ratio of t over the MTTF value.

Chapter 3

RELIABILITY AND SIX SIGMA MEASURES

If facts don't fit the theory, change the facts.
Albert Einstein

3.1 INTRODUCTION

In this chapter we discuss various measures by which reliability and Six Sigma characteristics can be numerically defined and described. Manufacturers use reliability and Six Sigma measures to quantify the product, production and process effectiveness. Use of any particular reliability and Six Sigma measure depends on what is expected of the system and what we are trying to measure. Several product development decisions are made using reliability and Six Sigma measures as one of the important design parameter. Technically, the term Six Sigma itself is a measure of quality. In this chapter we first summarize Six Sigma measurement system followed by reliability measures.

3.2 SIX SIGMA MEASURES

Processes make product and the effectiveness of the processes depend on its ability to satisfy customer requirements. Variability within the process leads to customer dissatisfaction. All the measures within Six Sigma measurement system attempt to quantify the capability of processes used in developing the product. Historically, process capability, C_p, and process capability index, C_{pk}, are used to measure the ability of the processes to meet the customer requirements. Process capability index and the process capability are defined in equation (3.1) and (3.2) respectively.

$$C_p = \frac{US - LS}{6\sigma} \tag{3.1}$$

$$C_{pk} = Min[(US - \mu)/3\sigma, (\mu - LS)/3\sigma] \tag{3.2}$$

Where, US and LS are upper and lower specifications (or lower and upper control limits, LCL and UCL) of the process respectively. Whereas, μ and σ stands for the process mean and standard deviation. A process capability ratio of 1 represents that the process will produce 99.73% of the products that meet the customer requirements. This also corresponds to the 3 sigma process quality (substituting US = μ + 3σ and LS = μ - 3σ in equation 3.1, we get C_p = 1). Six Sigma corresponds to the process capability ratio of 2 (substituting US = μ + 6σ and LS = μ - 6σ in equation 3.1, we get C_p = 2). In the following sections, we define various measurement systems used in Six Sigma.

3.3 YIELD

Yield, Y, is an important measure of process capability and is defined as the ratio of the total number of defect free units to the total number of units produced expressed in percentage. Mathematically, Yield can be written as:

$$Yield = Y = \frac{\text{Number of defect free units}}{\text{Number of opportunites}} \times 100\% \tag{3.3}$$

Alternatively,

$$Yield = Y = \frac{\text{Number of opportunities - Number of defects}}{\text{Number of opportunites}} \times 100\%$$

For example, if 456 units were produced and there were 18 defects, then Yield, Y, is given by:

$$Yield = \frac{456 - 18}{456} \times 100\% = \frac{438}{456} \times 100\% = 96.05\% \tag{3.4}$$

Table 3-1. Yield corresponding to different sigma levels

Sigma Level	Yield (Y)
6	99.999660
5	99.97670
4	99.3790
3	93.32
2	69.13
1	30.23

Table 3.1 gives the yield against various Sigma Level (with 1.5σ) shift. If the defect rate is known and the process is assumed to follow a discrete probability distribution, then one can use the probability mass function of that distribution to estimate the Yield of that process. For example, assume that process defects follow a Poisson process with defect rate $\lambda = 0.001$ (that is 1 defect per 1000 units), then the yield, Y, is given by the probability that the process will result in zero defect. That is, if the process defect is denoted by the random variable X, then, Yield, Y, is given by:

$$Y = P(X = 0) \times 100\% = \frac{e^{-\lambda} \times \lambda^{x}}{x!} \times 100\% = 99.90\% \tag{3.5}$$

3.4 DEFECTS PER MILLION OPPORTUNITIES (DPMO)

Defects per million opportunities (DPMO) measure the number of defects in a process in terms of million opportunities. It is also known as ppm (defects counted in parts per million). However, it is very difficult to get the complete data of defects in one million opportunities for most processes. Usually the DPMO is estimated from a sample of units produced. DPMO is calculated from estimating the Defects per Unit (DPU) using the following expression:

$$DPMO = DPU \times 10^{6} \tag{3.6}$$

Equation (3.6) can be written as:

$$DPMO = \frac{\text{Number of defects}}{\text{Number of opportunities}} \times 10^{6} \tag{3.7}$$

Assume that there were 348 units produced by a process and out of 348 units, 12 were classified as defective units after the quality test (DPU = 0.03448). Then, the DPMO, using equation (3.7), is given by:

$$DPMO = \frac{12}{348} \times 10^6 \approx 34483$$

For a given DPMO, the corresponding Yield is given by:

$$Yield = \left(1 - \frac{DPMO}{1000000}\right) \times 100\% \tag{3.8}$$

Table (3.2) gives the Yield corresponding to different DPMO values.

Table 3-2. Yield Corresponding to different DPMO values

DPMO	Yield
50	99.995%
500	99.95%
5000	99.5%
50000	95 %
500000	50 %

3.5 SIGMA QUALITY LEVEL

Sigma quality level is a measure of process defect rate. A higher Sigma level indicates that the process results in fewer defects, whereas a lower Sigma means higher defect rate. Sigma quality level can be used for benchmarking purpose and helps to measure quality of the process. Sigma quality also helps to set a realistic target for improvement of process quality during the DMAIC cycle of process improvement. Sigma level can be calculated from DPMO as well as from Yield.

3.5.1 Conversion of Yield to Sigma Level

Let Y denote yield of a process represented in percentage. Sigma level is nothing but the Z value of the standard normal distribution (if there is no shift in the process mean). There exist several polynomial approximations (Abramowitz and Stegun, 1972) to calculate Z value of the standard normal distribution. The value of Z actually gives the Sigma level when there is no shift in the process mean. The Sigma level under the assumption that the process mean is likely to shift as much as 1.5σ can be calculated using the following mathematical relationship:

$$\text{Sigma level} = 1.5 + \left(P - \frac{C_0 + C_1 P + C_2 P^2}{1 + d_1 P + d_2 P^2 + d_3 P^3} \right) \tag{3.9}$$

where,

$$P = \sqrt{\ln\left(\frac{1}{\left(1 - \frac{Y}{100}\right)^2} \right)} \tag{3.10}$$

$C_0 = 2.515517$, $C_1 = 0.802853$, $C_2 = 0.010328$, $d_1 = 1.432788$,

$d_2 = 0.189269$ and $d_3 = 0.001308$

$\left(P - \frac{C_0 + C_1 P + C_2 P^2}{1 + d_1 P + d_2 P^2 + d_3 P^3} \right)$ is an approximate expression for calculation of Z. In the Microsoft Excel® the function **NORMINV(Y/100, 1.5, 1)** gives the value of sigma level with 1.5σ shift. Not all process would shift by 1.5 σ, it is possible that the process mean may shift by less than 1.5 σ (and nothing stops it from shifting by more than 1.5 σ). In general, for a process with a process shift of Λ × σ, the Sigma level is given by:

$$\text{Sigma level} = \Lambda + \left(P - \frac{C_0 + C_1 P + C_2 P^2}{1 + d_1 P + d_2 P^2 + d_3 P^3} \right) \tag{3.11}$$

The corresponding Microsoft Excel® function is **NORMINV(Y/100,Λ, 1).**

EXAMPLE 3.1

A company involved in manufacturing car engine spare parts produced 80,000 units in the year 2004. 1290 unites were found to be defective out of 80,000 units. Calculate the process sigma level under 1.5 σ process shift.

The yield, Y, is given by:

$$Y = \frac{80000 - 1290}{80000} \times 100 = 98.3875,$$

From equation (3.10), the value of P is given by:

$$P = \sqrt{\ln\left(\frac{1}{\left(1 - \frac{Y}{100}\right)^2}\right)} = \sqrt{\ln\left(\frac{1}{\left(1 - \frac{98.3875}{100}\right)^2}\right)} = 2.8731$$

Using equation (3.9), we get the Sigma level as 3.6417.

3.5.2 Conversion of DPMO to Sigma Level

One can convert the DPMO to yield using equation (3.8), once the yield is known; the equations (3.9) – (3.10) can be used to calculate the Sigma level. Equation (3.12) can also be used to calculate approximate value of Sigma level under 1.5σ process shift (Schmidt and Launsby 1997; Bfreyfogle 2003):

$$\text{Sigma level} = 0.8406 + \sqrt{29.37 - 2.221 \times \ln(DPMO)} \tag{3.12}$$

Care should be taken while using equation (3.12) to calculate Sigma level. For DPMO value greater than 553365, the expression inside the square root function in equation (3.12) is negative. One can use the Excel® function **NORMINV(1 – DPMO/1000000, 1.5, 1)** to convert the DPMO to Sigma level.

EXAMPLE 3.2

DPMO of a process manufacturing car wind screen is 4000. Calculate the sigma level of this process. Calculate the Sigma level if the process shift is 0.5σ.

Using Excel function we get:

Sigma level = **NORMINV(1-4000/1000000, 1.5, 1)** = 4.1520

Alternatively,

$$\text{Sigma level} = 0.8406 + \sqrt{29.34 - 2.221 \times \ln(4000)} = 4.1449$$

If the process shift is 0.5 σ, the Sigma level is given by:

Sigma level = **NORMINV(1-4000/1000000, 0.5, 1)** = 3.1520

3.6 RELIABILITY MEASURES

The reliability measures used to specify reliability must reflect the real operational requirements of the product. Requirements must be tailored to individual products considering operational requirement and mission criticality. In broader sense, the reliability metrics can be classified (Figure 3.1) as: 1. Basic Reliability Measures, 2. Mission Reliability Measures, 3. Operational Reliability Measures, and 4. Contractual Reliability Measures.

Basic Reliability Measures are used to predict the system's ability to operate without maintenance and logistic support. Basic reliability measures try to measure the inherent strength of the product. How long a product can survive, or maintain the functionality without maintenance and support reflects the true strength of the product. Measures such as reliability function and failure function fall under this category.

Mission Reliability Measures are used to predict the system's ability to complete a mission. In many situations, it may be required to choose a product from a fleet to accomplish a mission. For example, in war scenario, the war strategists would like to know which aircraft from a fleet of aircraft should be deployed, so that the probabilities of survival of the chosen aircraft are maximized. Reliability measures such as mission reliability, maintenance free operating period (MFOP), failure free operating period (FFOP), and hazard function fall under this category.

Operational Reliability Measures are used to predict the performance of the system when operated in a planned environment including the combined effect of design, quality, environment, maintenance, support policy, etc. Measures such as Mean Time Between Maintenance (MTBM), Mean Time Between Overhaul (MTBO), Maintenance Free Operating Period (MFOP), Mean Time Between Critical Failure (MTBCF) and Mean Time Between Unscheduled Removal (MTBUR) fall under this category.

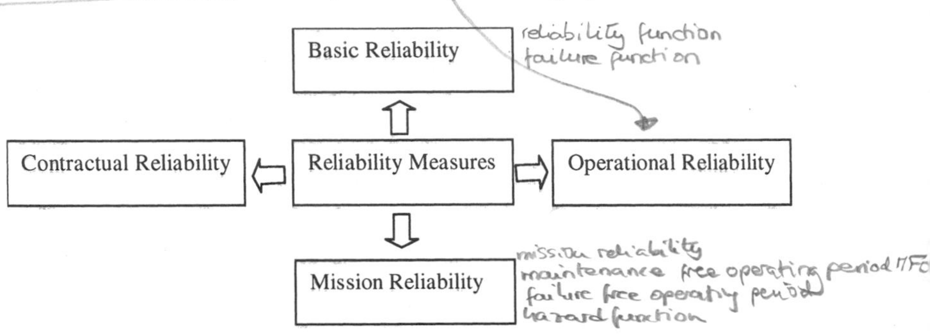

Figure 3-1. Classification of reliability measures

Contractual Reliability Measures are used to define measure and evaluate manufacturer's program. Contractual reliability is calculated by considering design and manufacturing characteristics. Measures such as Mean Time To Failure (MTTF), Mean Time Between Failure (MTBF) and Failure rate fall under this category.

Though we classify the reliability measures into four categories as mentioned above, one may require more than one reliability measure for specifying reliability requirements. This is due to the fact that complete information about the product reliability may not be captured with a single reliability measure. Selection of various measures to quantify the reliability requirements should include mission and logistic reliability along with maintenance and support measures. Currently, many manufacturers specify reliability using mean time between failure (MTBF) and failure rate. However, MTBF and failure rates have several drawbacks (which will be discussed later in this chapter). In the following sections, we define various reliability measures and how to calculate them. All the measures are defined based on the assumption that the time-to-failure (TTF) distribution of the system is known.

3.7 FAILURE FUNCTION

Failure function is a basic (logistic) reliability measure and is defined as the probability that an item will fail by time t. Here time t is used in a generic sense and it can have units such as miles, number of landings, flying hours, number of cycles, etc., depending on the operational profile and the utilization of the system. Failure function is equal to the probability that the time-to-failure random variable takes a value less than or equal to a particular value t (in this case operating time, see Figure 3.2). The failure function is usually represented as $F(t)$.

$$F(t) = \text{P (failure will occur before or at time t)} = \text{P } (TTF \leq t)$$

$$= \int_0^t f(u)du \tag{3.13}$$

Where $f(t)$ is the probability density function of the time-to-failure random variable *TTF*. Exponential, Weibull, normal, and lognormal are few popular theoretical distributions that are used to represent failure function. Equation (3.13) assumes that no maintenance is performed and gives the probability of failure by time t. However, most of the complex systems will require maintenance at frequent intervals. In such cases, equation (3.13) has to be modified, to incorporate the behaviour of the system under maintenance.

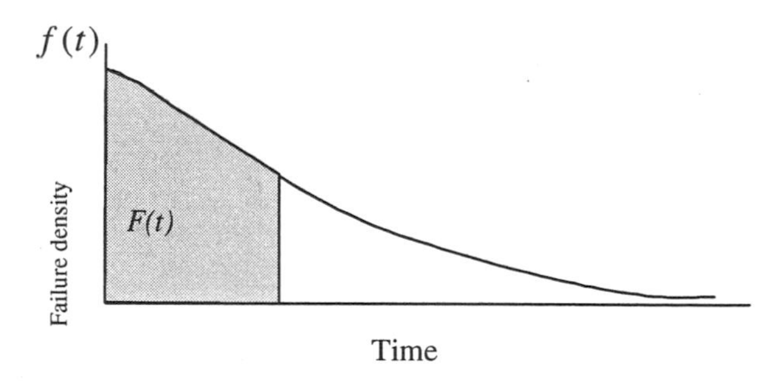

Figure 3-2. Failure function of a hypothetical distribution

Failure functions of few popular theoretical distributions are listed in Table 3.3. It should be noted that in the case of normal distribution the failure function exists between -∞ and +∞, so may have significant value at t ≤ 0. Since negative time is meaningless, great care should be taken in using normal distribution for the failure function. The best strategy is to use truncated normal distribution.

Table 3-3. Failure function, F(t), of few theoretical probability distributions

Distribution	Failure function F(t)
Exponential	$1-\exp(-\lambda t) \quad t>0, \lambda>0$
Normal	$\int_0^t \frac{1}{\sigma\sqrt{2\pi}} e^{-[\frac{1}{2}\left(\frac{x-\mu}{\sigma}\right)^2]} dx \quad or \quad \Phi\left(\frac{t-\mu}{\sigma}\right)$
Lognormal	$\int_0^t \frac{1}{\sigma_l x\sqrt{2\pi}} e^{-\left(\frac{1}{2}\left(\frac{\ln(x)-\mu_l}{\sigma_l}\right)^2\right)} dx \quad or \quad \Phi\left(\frac{\ln(t)-\mu_l}{\sigma_l}\right)$
Weibull	$1-\exp(-(\frac{t-\gamma}{\eta})^{\beta}) \quad \eta,\beta,\gamma>0, t\geq\gamma$

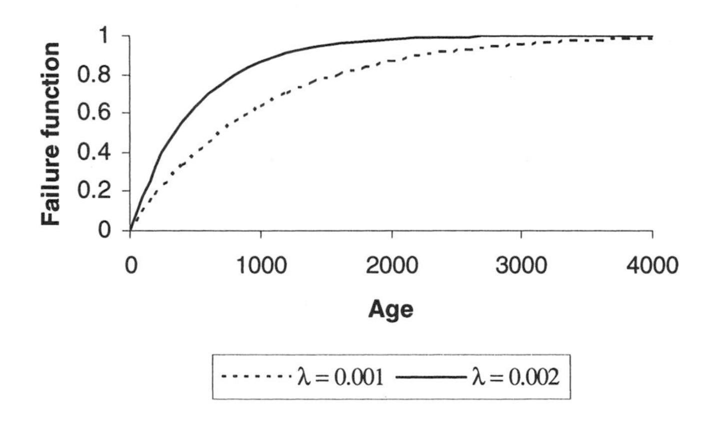

Figure 3-3. Failure function of exponential distribution for different parameter values

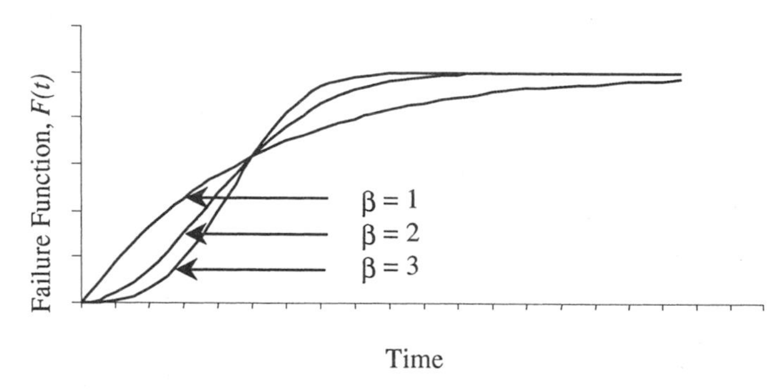

Figure 3-4. Failure function of Weibull distribution for different parameter values

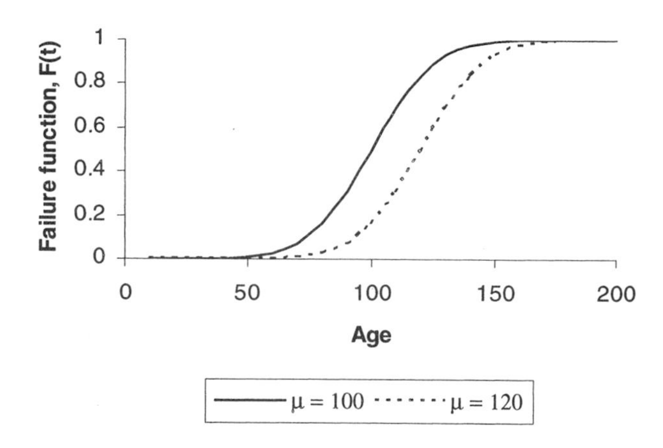

Figure 3-5. Failure function of normal distribution for different parameter values

Characteristics and applications of failure function are listed below. Failure functions of exponential, Weibull and normal distributions for different parameter values are shown in Figures 3.3 – 3.5.

Characteristics of failure function

1. Failure function is an increasing function, that is, for $t_1 < t_2$, $F(t_1) \leq F(t_2)$.

2. For modelling purposes it is assumed that the value of failure function at time t = 0, is 0 [F(0) = 0]. However, this assumption may not be valid always. For example, systems can be *dead on arrival*. The value of failure function increases as the time increases and for $t = \infty$, $F(\infty) = 1$.

Applications of failure function

1. $F(t)$ is the probability that an individual item will fail by time t.
2. $F(t)$ is the fraction of items that fail by time t.
3. 1 - $F(t)$ is the probability that an individual item will survive up to time t.

EXAMPLE 3.3

The time to failure distribution of a sub-system in an aircraft engine follows Weibull distribution with scale parameter η = 1100 flight hours and the shape parameter $\beta = 3$. Find:

a) Probability of failure during first 100 flight hours.
b) Find the maximum length of flight such that the failure probability is less than 0.05.

(a) Failure function of Weibull distribution is given by:

$$F(t) = 1 - \exp(\frac{t-\gamma}{\eta})^{\beta}$$

It is given that: t = 100 flight hours, η = 1100 flight hours, $\beta = 3$ and $\gamma = 0$. Probability of failure within first 100 hours is given by:

$$F(100) = 1 - \exp(-(\frac{100-0}{1100})^3) = 0.00075$$

b) If t is the maximum length of flight such that the failure probability is less than 0.05, we have

$$F(t) = 1 - \exp(-(\frac{t-0}{1100})^3) < 0.05$$

$$= \exp(-(\frac{t}{1100})^3) > 0.95$$

$$= (\frac{t}{1100})^3 < -\ln 0.95 \Rightarrow t = 1100 \times [-\ln(0.95)]^{1/3}$$

Now solving for t, we get t = 408.70 flight hours. The maximum length of flight such that the failure probability is less than 0.05 is 408.70 flight hours.

EXAMPLE 3.4

The time to failure distribution of a Radar Warning Receiver (RWR) system in a fighter aircraft follows Weibull distribution with scale parameter 1200 flight hours and shape parameter 3. The time to failure distribution of the same RWR in a helicopter follows exponential distribution with scale parameter 0.001. Compare the failure function of the RWR in the fighter aircraft and the helicopter. If the supplier gives a warranty for 750 flight hours, calculate the risk involved with respect to fighter aircraft and the helicopter. (Although we have a same system, the operating conditions have significant impact on the failure function. In this case, RWR in helicopter is subject to more vibrations compared to aircraft).

The failure function of RWR on the fighter aircraft is given by:

$$F(t) = 1 - \exp(-(\frac{t}{1200})^3)$$

The failure function of RWR on the helicopter is given by:

$$F(t) = 1 - \exp(-(0.001 \times t))$$

Figure 3.6 depicts the failure function of RWR in fighter aircraft and the helicopter. If the supplier provides warranty for 750 flight hours the risk associated with aircraft is given by:

$$F(750) = 1 - \exp(-(\frac{750}{1200})^3) = 0.2166$$

That is, just above 21% percent of RWR are likely to fail if the RWR is installed in the aircraft.

If the RWR is installed in helicopter then the associated risk is given by:

$$F(750) = 1 - \exp(-0.001 \times 750) = 0.5276$$

In the case of helicopter, more than 52% of the RWR's are likely to fail before the warranty period.

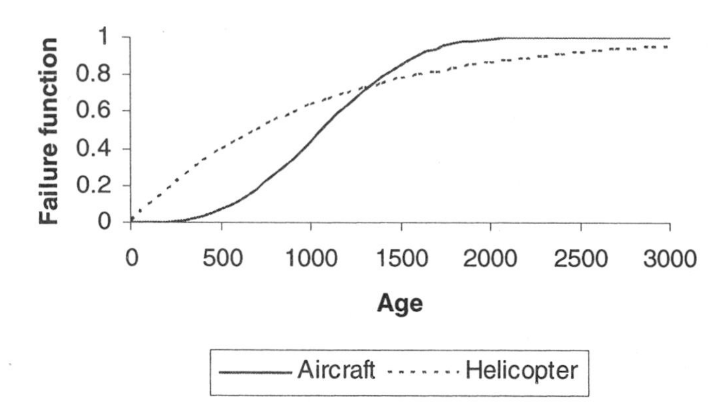

Figure 3-6. Failure function of RWR in fighter aircraft and helicopter

3.7.1 Failure function of systems under multiple failure mechanisms

It is seldom true that an item's failure is caused by a single failure mechanism. In most of the cases there will be more than one (some times hundreds) mechanism that causes the failure of an item. The expression for failure density (equivalent of equation 3.13) in case of multiple failure mechanisms can be derived if the time-to-failure distributions of various failure mechanisms are known. Assume that the system failure is caused due to two different failure mechanisms. Let $f_1(t)$ and $f_2(t)$ be the probability density function of the system due to failure mechanism 1 and 2 respectively. Now the probability density function of the time-to-failure of the system caused by either of the failure mechanisms is given by:

$$f(t) = f_1(t)[1 - F_2(t)] + f_2(t)[1 - F_1(t)]$$

Where, F_1(t) and F_2(t) are the failure functions for failure mechanism 1 and 2 respectively. The failure function of the item under two different failure mechanisms is given by:

$$F(t) = \int_0^t \{f_1(x)[1 - F_2(x)] + f_2(x)[1 - F_1(x)]\}dx \tag{3.14}$$

EXAMPLE 3.5

Failure of an item is caused by two different failure mechanisms (say failure mechanism A and B). The time-to-failure distribution of the item due to failure mechanism A can be represented by exponential distribution with parameter $\lambda_A = 0.002$ hours. The time-to-failure distribution of the item due to failure mechanism B can be represented by exponential distribution with parameter $\lambda_B = 0.005$ hours. Find the probability that the item will fail before 500 hours of operation.

Assume that $f_A(t)$ and $f_B(t)$ represent probability density functions of the time-to-failure random variable due to failure mechanisms A and B respectively. Thus,

$$f_A(t) = \lambda_A \exp(-\lambda_A t), \quad 1 - F_A(t) = \exp(-\lambda_A t)$$
$$f_B(t) = \lambda_B \exp(-\lambda_B t), \quad 1 - F_B(t) = \exp(-\lambda_B t)$$

Now the failure function of the item is given by:

$$\begin{aligned} F(t) &= \int_0^t \{\lambda_A \exp(-(\lambda_A + \lambda_B)x) + \lambda_B \exp(-(\lambda_A + \lambda_B)x)dx \\ &= (\lambda_A / \lambda_A + \lambda_B)[1 - \exp(-(\lambda_A + \lambda_B)t] \\ &\quad + (\lambda_B / \lambda_A + \lambda_B)[1 - \exp(-(\lambda_A + \lambda_B)t] \\ &= [1 - \exp(-(\lambda_A + \lambda_B)t] \end{aligned}$$

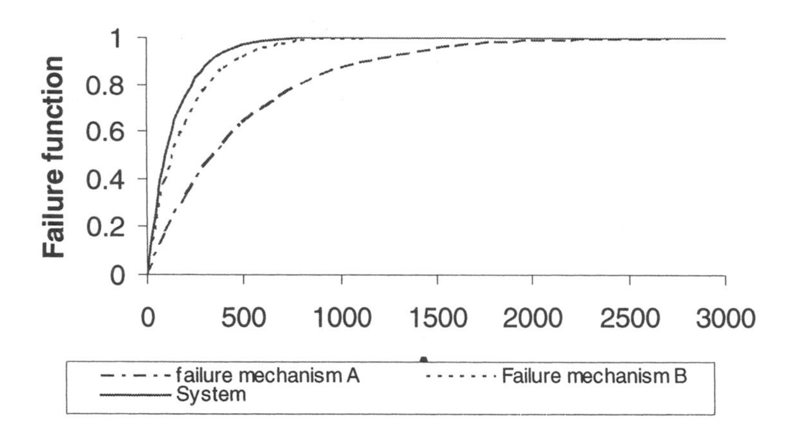

Figure 3-7. Failure function under more than one failure mechanisms

Figure 3.7 represents the failure function due to failure mechanism 1, 2 and the system failure function. The probability that the item will fail by 500 hours is given by:

$$F(500) = 1 - \exp(-((0.005 + 0.002) \times 500)) = 0.9698$$

3.8 RELIABILITY FUNCTION

Reliability is the ability of a product or an item to maintain the required function for a specified period of time (or mission time) under given operating conditions. *Reliability function, $R(t)$, is defined as the probability that the system will not fail during the stated period of time, t, under stated operating conditions.* If *TTF* represents the time-to-failure random variable with failure function (cumulative distribution function) $F(t)$, then the reliability function $R(t)$ is given by:

$$R(t) = \text{P}\{\text{the system doesn't fail during } [0\ , \text{t}]\} = 1 - F(t) \qquad (3.15)$$

In equation (3.15) we assume that the age of the system before the start of the mission is zero. Thus the equation (3.15) is valid only for new systems or those systems whose failures are not age related (that is, the time-to-failure follows exponential distribution). However, in most of the cases this assumption may not be valid.

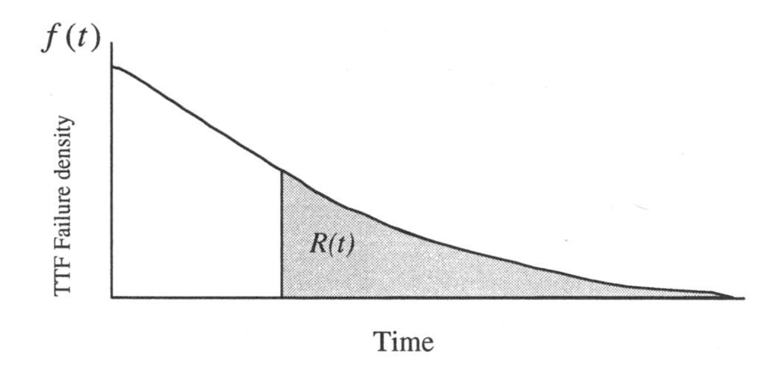

Figure 3-8. Reliability Function

If the system age is greater than zero at the beginning of the mission, then we have to calculate mission reliability function, which will be discussed later in this chapter. Figure 3.8 shows the relation between reliability function and the *TTF* density function. *R(t)* is the area under *TTF* density between t and ∞.

Properties of reliability function:

1. Reliability is a decreasing function with time t, that is, for $t_1 < t_2$; $R(t_1) \geq R(t_2)$.
2. It is usually assumed that R (0) = 1. As the value of t increases R(t) approaches zero, that is, $R(\infty) = 0$.

Applications of reliability function

1. R(t) is the probability that an individual item survives up to time t.
2. R(t) is the fraction of items in a population that survive up to time t.
3. R(t) is the basic function used for many reliability measures and system reliability prediction.

Reliability functions for some important life distributions are given in Table 3.4. Figures 3.9 – 3.11 represent reliability functions of various theoretical distributions for different parameter values.

Table 3-4. Reliability function of different time-to-failure distributions

Time-to-failure distribution	Reliability function
Exponential	$\exp(-\lambda t) \qquad t>0, \lambda>0$
Normal	$\Phi(\frac{\mu-t}{\sigma}) = 1-\int_0^t \frac{1}{\sigma\sqrt{2\pi}} e^{-\left(\frac{1}{2}\left(\frac{x-\mu}{\sigma}\right)^2\right)} dx$
Weibull	$\exp(-(\frac{t-\gamma}{\eta})^{\beta}) \qquad \eta, \beta, \gamma > 0, t \geq \gamma$
Log-normal	$\Phi(\frac{\mu_l - \ln t}{\sigma_l}) = 1-\int_0^t \frac{1}{\sigma_l x\sqrt{2\pi}} e^{-\left(\frac{1}{2}\left(\frac{\ln(x)-\mu_l}{\sigma_l}\right)^2\right)} dx$

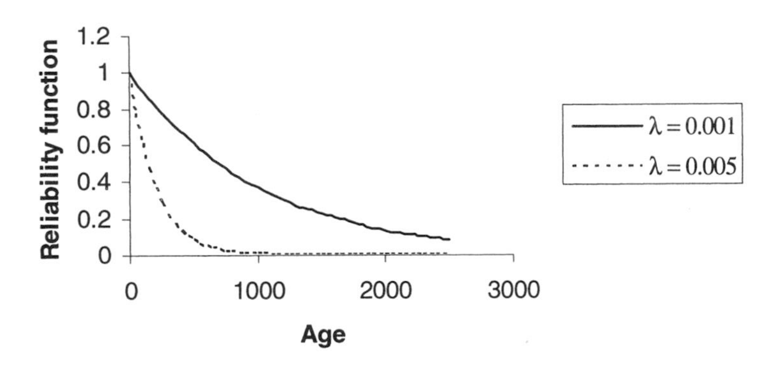

Figure 3-9. Reliability function of exponential distribution

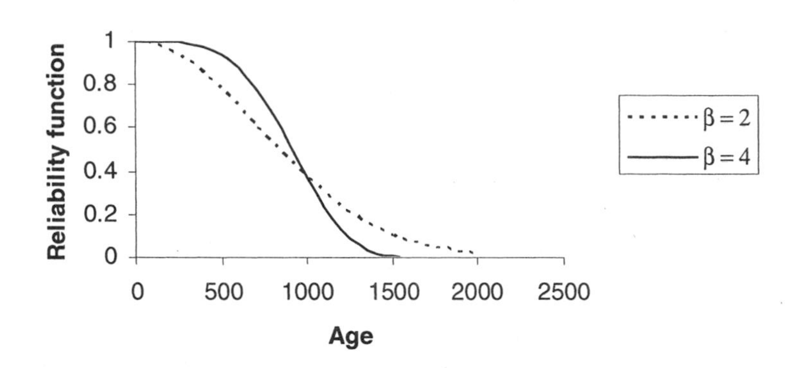

Figure 3-10. Reliability function of Weibull distribution for different values of β and η = 1000

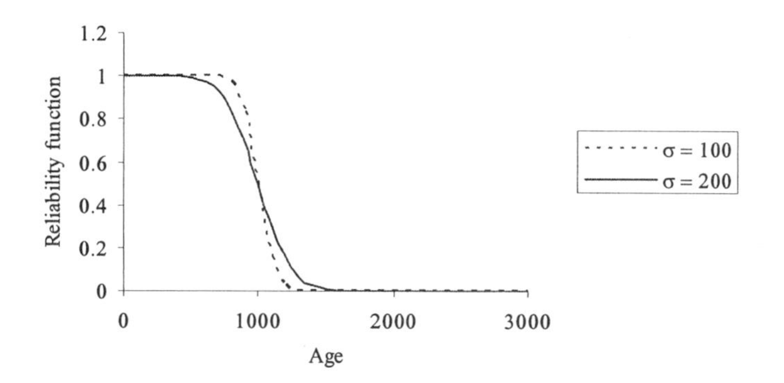

Figure 3-11. Reliability function of normal distribution for different values of σ

EXAMPLE 3.6

Time to failure distribution of a computer memory chip follows normal distribution with mean 9000 hours and standard deviation 2000 hours. Find the reliability of this chip for a mission of 8000 hours.

Using Table 3.4, the reliability for a mission of 8000 hours is given by:

$$R(t) = \Phi(\frac{\mu - t}{\sigma}) = \Phi(\frac{9000 - 8000}{2000}) = \Phi(0.5) = 0.6915$$

EXAMPLE 3.7

The time to failure distribution of a steam turbo generator can be represented using Weibull distribution with η = 500 hours and β = 2.1. Find the reliability of the generator for 600 hours of operation.

From Table 3.4, reliability of the generator for 600 hours of operations is given by:

$$R(t) = \exp(-(600/500)^{2.1}) = 0.2307$$

3.8.1 Reliability function of items under multiple failure mechanisms

Assume that the failure of the item is caused due to two different failure mechanisms. Let $f_1(t)$ and $f_2(t)$ be the probability density functions of the time-to-failure random variable due to failure mechanisms 1 and 2 respectively. The probability density function of the time-to-failure of the item caused by either of the failure mechanisms is given by:

$$f(t) = f_1(t)[1 - F_2(t)] + f_2(t) \cdot [1 - F_1(t)]$$

Where F_1(t) and F_2(t) are failure functions for failure mechanisms 1 and 2 respectively. The Reliability function of the item under two different failure mechanisms is given by:

$$R(t) = 1 - F(t) = 1 - \int_0^t \{f_1(x)[1 - F_2(x)] + f_2(x)[1 - F_1(x)]\}dx \quad (3.16)$$

The above result can be extended to obtain expression for reliability function due to more than two failure mechanisms.

EXAMPLE 3.8

For the system described in example 3.5, find the reliability of the item for 200 hours.

Using the expression for failure function obtained in example 3.5, the reliability function can be written as:

$$R(t) = \exp(-(\lambda_A + \lambda_B) \times t)$$
$$R(200) = \exp(-(0.002 + 0.005) \times 200) = 0.2465$$

3.8.2 Reliability function and Six Sigma

Six Sigma strategy was proposed to improve the reliability of the products developed by Motorola. In this section, we develop models to integrate the Six Sigma metric, Sigma level quality, and the reliability function. Reliability target is set during the design stage of any product and not meeting the target reliability is likely to result in customer dissatisfaction. The defect in this case is the additional number of failures that a customer is likely to face due to not meeting the reliability target. Failure to meet target reliability can be attributed to the capability of the processes associated with design and development of that product. The Sigma level of a process corresponding to meeting the reliability target can be derived by calculating the difference between the target reliability, $R_g(t)$, and the achieved reliability, R_a (t). The defects per unit (DPU) for not meeting the target reliability is given by:

$$DPU = R_g(t) - R_a(t), \quad \mathrm{R_a}(t) < R_g(t)$$

The corresponding DPMO is given by:

$$DPMO = [R_g(t) - R_a(t)] \times 10^6 \qquad (3.17)$$

Equation (3.17) gives the additional number of failures one can expect in a fleet of one million units by time t for not meeting the target reliability. The sigma level can be calculated using equation (3.9), where the value of P is given by:

$$P = \sqrt{\ln\left(\frac{1}{(1-[1-(R_g(t)-R_a(t)])^2}\right)} = \sqrt{\ln\left(\frac{1}{[(R_g(t)-R_a(t)]^2}\right)} \tag{3.18}$$

The corresponding Excel® function for calculation of Sigma level is given by:

NORMINV(1 – [R_g(t) – R_a(t)] , 1.5, 1) (3.19)

EXAMPLE 3.9

During the development of a battle tank, a target reliability of 0.85 was set for 1000 kilometers of usage. During the operation, it was observed that the achieved reliability was 0.8447 for 1000 miles usage. Calculate the Sigma level of the development process.

The DPU is given by:

$$DPU = R_g(t) - R_a(t) = 0.85 - 0.84447 = 0.00553$$

The corresponding DPMO is 5530. Sigma level can be calculated using Excel function defined in (3.19).

Sigma level = NORMINV(1-0.00553, 1.5, 1) = 4.0407

3.8.3 Mission Reliability

In many practical situations, one might be interested in finding the probability of completing a mission successfully. For a fighter aircraft, the success probability of hitting an enemy target and returning to the base is an example where mission reliability function can be used. The main difference between reliability function and the mission reliability function is that, in mission reliability we recognize the age of the system before the mission. *Mission reliability is defined, as the probability that the system aged t_b is able to complete mission of duration t_m successfully.* We assume that no maintenance is performed during the mission. The expression for mission reliability MR (t_b , t_m) is given by

$$MR(t_b, t_m) = \frac{R(t_b + t_m)}{R(t_b)} \tag{3.20}$$

where, t_b is the age of the item at the beginning of the mission and t_m is the mission period. If the time to failure distribution is exponential, then the following relation is valid.

$$MR(t_b, t_m) = R(t_m)$$

Application of mission reliability function

1. Mission reliability, $MR(t_a, t_m)$ gives the probability that an individual item aged t_a will complete a mission duration of t_m hours without any need for maintenance.
2. Mission reliability is the appropriate basic reliability measure for ageing items or items whose time-to-failure distribution is not exponential.

EXAMPLE 3.10

Time-to-failure distribution of the gearbox within an armoured vehicle can be modeled using Weibull distribution with scale parameter $\eta = 2400$ miles and shape parameter $\beta = 1.25$. Find the probability that that gearbox will not fail during a mission time of 200 miles, assuming that the current age of the gearbox is 1500 miles.

Given, $t_b = 1500$ miles and $t_m = 200$ miles

$$MR(t_b, t_m) = \frac{R(t_m + t_b)}{R(t_b)} = \frac{R(1700)}{R(1500)}$$

$$R(1700) = \exp(-(\frac{1700}{2400})^{1.25}) = 0.5221$$

$$R(1500) = \exp(-(\frac{1500}{2400})^{1.25}) = 0.5736$$

$$MR(1500,200) = \frac{R(1700)}{R(1500)} = \frac{0.5221}{0.5736} = 0.9102$$

That is, the gearbox aged 1500 miles has approximately 91% chance of surviving a mission of 200 miles.

3.9 HAZARD FUNCTION (INSTANTANEOUS FAILURE RATE)

Hazard function (or hazard rate) is used as a parameter for comparison of two different designs in reliability theory. Hazard function is the indicator of the impact of ageing on the reliability of the system. It quantifies the risk of failure as the age of the system increases. Mathematically, it represents the limiting value of the conditional probability of failure in an interval t to t + δt given that the system survives up to t, divided by δt, as δt tends to zero, that is,

$$h(t) = \lim_{\delta t \to 0} \frac{1}{\delta t} \cdot \frac{F(t+\delta t) - F(t)}{R(t)} = \lim_{\delta t \to 0} \frac{R(t) - R(t+\delta t)}{\delta t R(t)} \tag{3.21}$$

Note that hazard function, h(t), is not a probability, it is the limiting value of the probability. However, h(t)δt, represents the probability that the item will fail between ages t and t+δt as δt →0. The above expression can be simplified as follows:

$$h(t) = \frac{f(t)}{R(t)} \tag{3.22}$$

Thus, the hazard function is the ratio of the probability density function to the reliability function. Integrating both sides of the above equation, we get:

$$\int_0^t h(x)dx = \int_0^t \frac{f(x)}{R(x)} dx$$

$$= \int_0^t -\frac{R'(x)}{R(x)} dx = -\ln R(t)$$

Thus reliability can be written as:

$$R(t) = \exp\left[-\int_0^t h(x)dx \right] \tag{3.23}$$

From equation (3.23), it immediately follows that:

$$f(t) = h(t)\exp(-\int_0^t h(x)dx \tag{3.24}$$

The expression (3.24), which relates reliability and hazard function, is valid for all types of time to failure distributions. Hazard function shows how the risk of the item in use changes over time (hence also called *risk rate).* The hazard functions of some important theoretical distributions are given in Table 3.5.

Characteristics of hazard function

1. Hazard function can be increasing, decreasing or constant.
2. Hazard function is not a probability and hence can be greater than 1.

Applications of hazard function

1. h(t) is casually referred as failure rate at time t (time-dependent)
2. h(t) quantifies the amount of risk a system is under at time t.
3. For h(t) $\leq$ 1, it is not recommended to carry out preventive maintenance.

Figures 3.12-3.14 show hazard functions of various theoretical distributions for different parameter values.

Table 3-5. Hazard function of different distributions

Distribution	Hazard function
Exponential	λ
Weibull	$\frac{\beta}{\eta}(\frac{t}{\eta})^{\beta-1}$
Normal	$f(t)/\Phi(\frac{\mu-t}{\sigma})$, *f(t)* is the pdf of normal distribution
Log-normal	$f_l(t)/\Phi(\frac{\mu_l-t}{\sigma_l})$, $f_l(t)$ is the pdf of lognormal distribution.

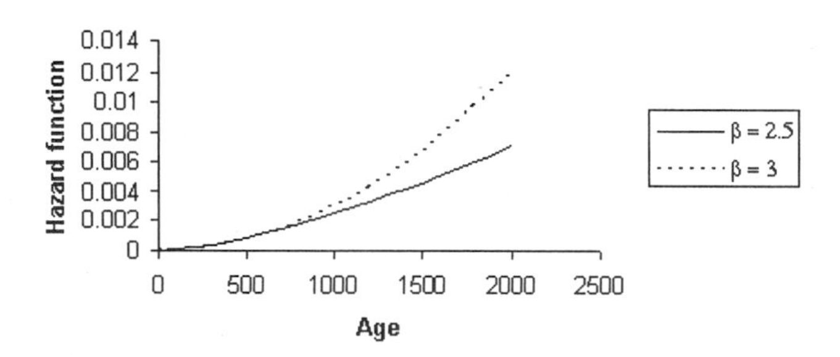

Figure 3-12. Hazard function of Weibull distribution for different values of β

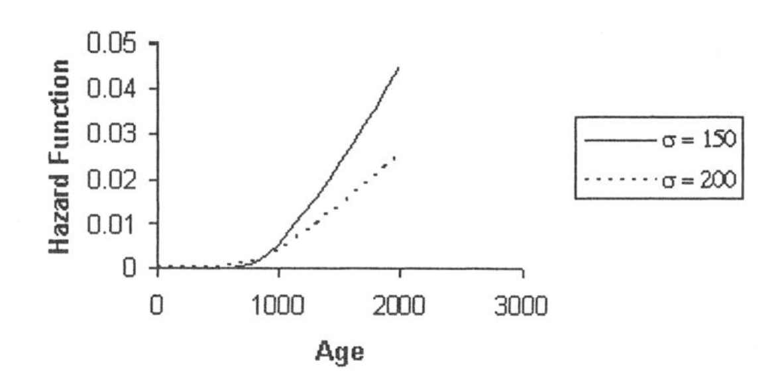

Figure 3-13. Hazard function of normal distribution for different values of σ

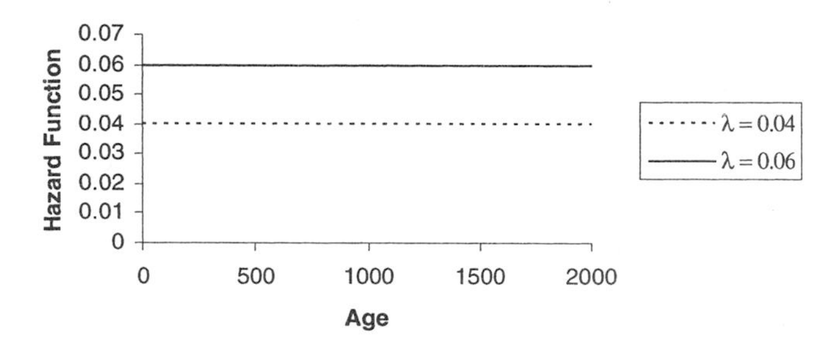

Figure 3-14. Hazard function of exponential distribution for different values of λ

EXAMPLE 3.11

Time to failure distribution of a gas turbine system can be represented using Weibull distribution with scale parameter η = 1000 hours and shape parameter β = 1.7. Find the hazard rate of the gas turbine at time t = 800 hours and t = 1200 hours.

The hazard rate for Weibull distribution is given by:

$$h(t) = \frac{\beta}{\eta}\left(\frac{t}{\eta}\right)^{\beta-1}$$

$$h(800) = \frac{1.7}{1000}(\frac{800}{1000})^{0.7} = 0.00145$$

$$h(1200) = \frac{1.7}{1000}(\frac{1200}{1000})^{0.7} = 0.0019$$

3.9.1 Cumulative hazard function

Cumulative hazard function represents the cumulative hazard or risk of the item during the interval [0,t]. Cumulative hazard function, H(t), is given by:

$$H(t) = \int_0^t h(x)dx \tag{3.25}$$

Reliability of an item can be conveniently written using cumulative hazard as:

$$R(t) = e^{-H(t)} \tag{3.26}$$

Consider an item, which upon failure is subject to minimal repair. That is, the hazard rate after repair is same as the hazard rate just before failure. If $N(t)$ is the total number of failures by time t, then, $M(t) = E[N(t)]$, is the expected number of failures by time t. It can be shown under the assumption that the item receives minimal repair* (*'as-bad -as-old'*) after each failure that,

$$E[N(t)] = M(t) = \int_0^t h(x)dx \tag{3.27}$$

The above expression can be used to model different maintenance/replacement policies. In case of exponential and Weibull time

* Mathematically minimal repair or *'as bad as old'* means that the hazard rate of the item after repair will be same as the hazard rate just prior to failure.

to failure distributions we get the following simple expressions for the expected number of failures of an item subject to minimal repair.

Exponential time to failure distribution

For exponential distribution, the expected number of failures is given by

$$E[N(t)] = \int_0^t h(x)dx = \int_0^t \lambda dx = \lambda t \tag{3.28}$$

Weibull time to failure distribution

For Weibull distribution, the expected number of failures is given by

$$E[N(t)] = \int_0^t h(x)dx = \int_0^t \frac{\beta}{\eta}\left(\frac{x}{\eta}\right)^{\beta-1} dx = \left(\frac{t}{\eta}\right)^{\beta} \tag{3.29}$$

In practice, hazard function can have different shapes. Figure 3.15 shows most general forms of hazard function. Recent research in the field of reliability centered maintenance (RCM) shows that the hazard rate mostly follows six different patterns. Depending on the equipment and its failure mechanism, one can say that the hazard function may follow any one of these six patterns. *However, one should not blindly assume that hazard rate of any item will follow any one of these six patterns.* These are only possible cases based on some data.

Pattern A is called the *bathtub curve* and consist of three distinct phases. It starts with early failure region (known as burn-in or infant mortality) characterized by decreasing hazard function. Early failure region is followed by constant or gradually increasing region (called useful life). The constant or gradually increasing region is followed by wear out region characterized by increasing hazard function. The reason for such as shape is that the early decreasing hazard rate results from manufacturing defects. Early operation will remove these items from a population of like items. The remaining items have a constant hazard for some extended period of time during which the failure cause is not readily apparent. Finally those items remaining reach a wear-out stage with an increasing hazard rate. One would expect bathtub curve at the system level and not at the part or component level (unless the component has many failure modes which have different *TTF* distribution). It was believed that bathtub curve represents the most general form of the hazard function. However, the recent research shows that in most of the cases hazard function do not follow this pattern.

Pattern B starts with high infant mortality and then follows a constant or very slowly increasing hazard function. Pattern C starts with a constant or slowly increasing failure probability followed by wear out (sharply increasing) hazard function. Pattern D shows constant hazard throughout the life. Pattern E represents a slowly increasing hazard without any sign of wear out. Pattern F starts with a low hazard initially followed by a constant hazard. These six patterns denote most frequently occurring forms of hazard function, in reality, the hazard function of a system can take any form depending on the probability density and reliability function of that system. Table 3.6 shows the relationship between failure function, reliability function and hazard function.

Table 3-6. Relationship between reliability function, failure function and hazard function

	F(t)	R(t)	h(t)
F(t)	-----	1 – R(t)	$1-\exp(-\int_0^t h(x)dx)$
R(t)	1 – F(t)	-----	$\exp(-\int_0^t h(x)dx$
h(t)	$F^{'}(t)/[1-R(t)]$	$-R^{'}(t)/R(t)$	-----

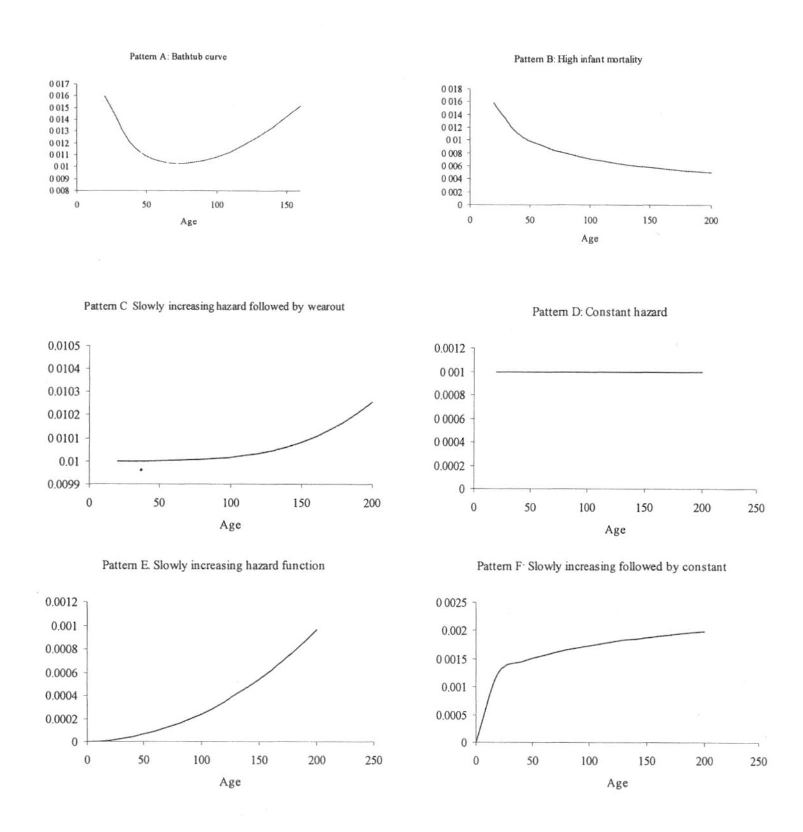

Figure 3-15. Different patterns of hazard function

EXAMPLE 3.12

An item is subjected to minimal repair whenever it failed. If the time to failure of the item follows Weibull distribution with $\eta = 500$ and $\beta = 2$. Find: 1. the number of times the item is expected to fail by 1500 hours, and 2. if the cost of the item is \$ 200, and the cost of minimal repair is \$ 100 per each repair, is it advisable to repair or replace the item upon failure.

1. The expected number of failures is given by:

$$E[N(t)] = [\frac{t}{\eta}]^{\beta} = [\frac{1500}{500}]^2 = 3^2 = 9$$

2. Using the above result the cost associated with repair, C_{repair} (t) = 9 × 100 = $ 900. If the item is replaced, then the expected number of failures is given by the renewal function, *M(t),* where

$$M(t) = \sum_{i=1}^{\infty} F^i(t)$$

For the above case, the value of M(t) < 4 (Calculation of the above function is discussed in Chapter 2), and hence the cost due to replacement will be less than 4 × 200 = $ 800. Thus, it is better to replace the item upon failure rather using minimal repair.

3.9.2 Failure rate

Whenever the hazard function is constant, we call it as failure rate. That is, failure rate is a special case of hazard function (hazard function is time dependent failure rate). Failure rate is one of the most widely used contractual reliability measures in practice. However, it is appropriate to use failure rate only when the time-to-failure distribution is exponential. Also, failure rate can be used only for non-repairable systems. Many defence standards such as MIL-HDBK-217 and British standard DEF-STAN 00-40 recommend the following equation for estimating the failure rate.

$$Failure\ rate = \frac{Total\ number\ of\ failures\ in\ a\ sample}{Cumulative\ operating\ time\ of\ the\ sample} \quad (3.30)$$

Care should be taken in using the above equation, for good estimation one has to observe the system failure for a sufficiently large operating period.

Applications of failure rate

1. Failure rate represents the number of failures per unit time. However, it is normal practice to quote failure rate for some specified duration of operation. For example, the failure rate (removal rate to be precise) of engines is quoted for 1000 flying hours.

2. If the failure rate is λ, then the expected number of items that fail in [0,t] is λt, since the time to failure is exponential, the failure process is Poisson with mean λt.
3. Failure rate is one of the popular contractual reliability measures among many industries including aerospace and defence.

3.10 MEAN TIME TO FAILURE

MTTF represents the expected value of a system's time to failure random variable. It is used as a measure of reliability for *non-repairable* items such as bulb, microchips and many electronic circuits. Mathematically, MTTF can be defined as:

$$MTTF = \int_0^\infty tf(t)dt = \int_0^\infty R(t)dt \tag{3.31}$$

Thus, MTTF can be considered as the area under the curve represented by the reliability function, R(t), between zero and infinity. If the item under consideration is repairable, then the expression (3.31) represents mean time to first failure of the item. Figure 3.16 depicts the MTTF of an item.

For many reliability functions, it is difficult to evaluate the integral (3.31). One may have to use numerical approximation such as trapezium approach to find MTTF value.

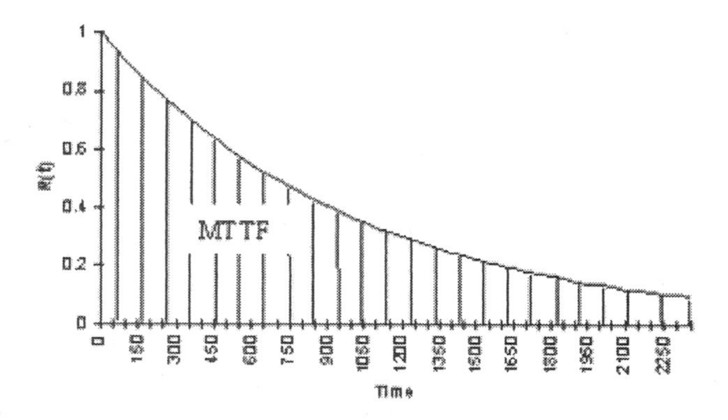

Figure 3-16. MTTF as a function of reliability

MTTF is one of the most popular measures for specifying reliability of non-repairable items among military and Government organizations throughout the world. Unfortunately there are many misconceptions about MTTF among reliability analysts. During the 1991 *Gulf War,* one of Generals from a defence department said, '*We know exactly how many tanks to send; we measured the distance from the map and divided that by MTTF'.* What many people do not realize is that MTTF is only a measure of central tendency. For example, if the time-to-failure distribution is exponential, then 63% of the items will fail before their age reaches MTTF value.

MTTF is one of the important contractual reliability measures for non-repairable (consumable) items. However, it is important to understand what MTTF value really means. For example let us assume that we have two items A and B with same MTTF (say 500 days). One might think that both the components have equal reliability. However, if the time to failure of the item A is exponential and that of item B is normal then there will be a significant variation in the behaviour of these items. Figure 3.17 shows the cumulative distribution of these two items up to 500 days. The figure clearly shows that items with exponential failure time show higher chance of failure during the initial stages of operation. It is easy to check that if the time to failure of the item is exponential then more than 63% of the items will fail by the time the age of the item reaches MTTF. In the case of normal distribution, it will be 50%.

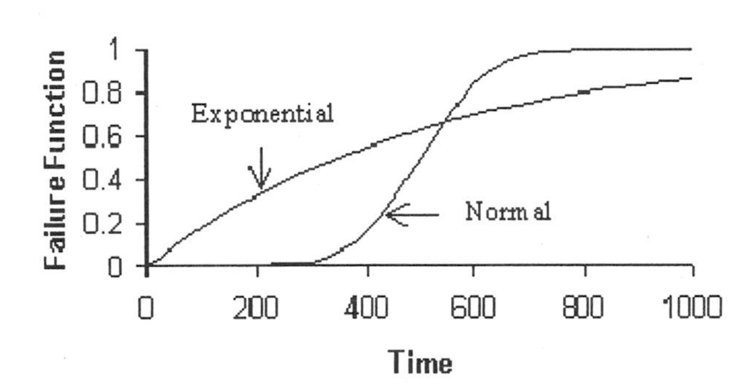

Figure 3-17. Reliability function of items with same MTTF

Using the equation (3.31), the MTTF of various failure distributions can be derived and are listed in Table 3.7.

Applications of MTTF

1. MTTF is the average life of a non-repairable system.
2. For a repairable system, MTTF represents the average time before the first failure.
3. MTTF is one of the popular contractual reliability measures for non-repairable systems.

Table 3-7. MTTF of different time-to-failure distributions

Distribution	Mean Time To Failure
Exponential	$1/\lambda$
Weibull	$\eta \times \Gamma(1+\frac{1}{\beta})$
Normal	μ
Log-normal	$\exp(\mu_l + \frac{\sigma_l}{2})$

3.10.1 MTTF and Sigma level

Mean Time to Failure (or its inverse failure rate) is the most popular reliability measure among practitioners. The customer requirement for target reliability of product is usually specified using MTTF. If $MTTF_t$ and $MTTF_a$ denote the target and achieved mean time to failure values respectively, the DPU for not meeting the target $MTTF_t$ is given by:

$$DPU = \left[\frac{1}{MTTF_a} - \frac{1}{MTTF_t}\right], \quad MTTF_a < MTTF_t \tag{3.32}$$

Equation (3.32) gives the additional number of failure one can expect per unit time for not meeting the target MTTF. The corresponding DPMO is given by:

$$DPMO = [\frac{1}{MTTF_a} - \frac{1}{MTTF_t}] \times 10^6 \tag{3.33}$$

Equation (3.33) gives the additional number of failures one can expect due to not meeting the target MTTF if the system is used for one million life units. The sigma level can be calculated using equation (3.9), where the value of P is given by:

$$P = \sqrt{\ln\left(\frac{1}{\left(\frac{1}{MTTF_a} - \frac{1}{MTTF_t}\right)^2}\right)} \tag{3.34}$$

The Excel® function for calculating the sigma level is given by:

$$\text{NORMINV}\left(1 - \left(\frac{1}{\text{MTTF}_\text{a}} - \frac{1}{\text{MTTF}_\text{t}}\right), 1.5, 1\right) \tag{3.35}$$

EXAMPLE 3.13

Powerplus, a battery manufacturer was given a target mean time to failure of 2000 hours by its main customer for their new maintenance free batteries. However, after sufficient data collection it was found that the achieved mean time to failure of batteries was 1950 hours. Calculate the sigma level of the processes used in manufacturing the maintenance free batteries.

Using Excel® function given in equation (3.35), the sigma level is given by:

$$\text{NORMINV}\left(1 - \left(\frac{1}{1950} - \frac{1}{2000}\right), 1.5, 1\right) = 5.7092$$

3.10.2 Mean Residual Life

In some cases, it may be of interest to know the expected value of the remaining life of the item before it fails from an arbitrary time t_0 (known as, *mean residual life*). We denote this value as $MTTF(t_0)$, which represents the expected time to failure of an item aged t_0. Mathematically, $MTTF(t_0)$ can be expressed as:

$$MTTF(t_0) = \int_{t_0}^{\infty} (t - t_0) f(t|t_0) dt \tag{3.36}$$

$f(t|t_0)$ is the density of the conditional probability of failure at time t, provided that the item has survived up to time t_0. Thus,

$$f(t \mid t_0) = h(t) \times R(t \mid t_0)$$

Where, $R(t|t_0)$, is the conditional probability that the item survives up to time t, given that it has survived up to time t_0. Now, the above expression can be written as:

$$f(t \mid t_0) = h(t) \times \frac{R(t)}{R(t_0)}$$

The expression for $MTTF(t_0)$ can be written as:

$$MTTF(t_0) = \int_{t_0}^{\infty} (t - t_0) h(t) \frac{R(t)}{R(t_0)} dt \tag{3.37}$$

substituting for h(t) in equation (3.37), we have

$$MTTF(t_0) = \int_{t_0}^{\infty} \frac{(t - t_0) f(t)}{R(t_0)} dt = \frac{1}{R(t_0)} \int_{t_0}^{\infty} (t - t_0) f(t) dt$$

The above equation can be written as (using integration by parts):

$$MTTF(t_0) = \frac{\int_{t_0}^{\infty} R(t)dt}{R(t_0)} \tag{3.38}$$

The concept of mean residual life can be successfully applied for planning maintenance and inspection activities.

EXAMPLE 3.14

Companies A and B manufacture car tyres. Both the companies claim that the MTTF of their car tyre is 2000 miles. After analysing the field failure data of these two tyres it was found that the time to failure distribution of A is exponential with λ = 0.0005 and the time to failure distribution of B is normal with μ = 2000 miles and σ = 200 miles. If the maintenance policy of the Hoboken city car rentals is to replace the tyres as soon as it reaches 2000 miles which tyre they should buy:

Reliability of the car tyre produced by company A for 2000 miles, $R_A(2000)$, is given by:

$$R_A(2000) = \exp(-0.0005 \times 2000) = 0.3678$$

Reliability of the car tyre produced by company B for 2000 miles, $R_B(2000)$, is given by:

$$R_B(2000) = \Phi(\frac{\mu - 2000}{\sigma}) = \Phi(\frac{2000 - 2000}{200}) = \Phi(0) = 0.5$$

Thus, it is advisable to buy the tyres produced by company B.

EXAMPLE 3.15

The time to failure of an airborne navigation radar can be represented using Weibull distribution with scale parameter η = 2000 hours and β = 2.1. It was told that the age of the existing radar is 800 hours. Find the expected value of the remaining life for this radar.

Using Equation (3.38), The MTTF(800) can be written as:

$$MTTF(800) = \frac{\int_{800}^{\infty} R(t)dt}{R(800)} = \frac{\int_{0}^{\infty} R(t)dt - \int_{0}^{800} R(t)dt}{R(800)}$$

$$\text{MTTF(800)} = MTTF(800) = \frac{MTTF - \int_{0}^{800} \exp(-(\frac{t}{2000})^{2.1} dt}{0.8641}$$

$$MTTF = \eta \times \Gamma(1+\frac{1}{\beta}) = 2000 \cdot \Gamma(1+\frac{1}{2.1}) = 1771.2$$

The value of $\Gamma(1+\frac{1}{\beta})$ can be found from Gamma function table (see appendix).

Using numerical approximation, $\int_{0}^{800} \exp(-(\frac{t}{2000})^{2.1} dt \approx 763.90$

Thus MTTF(800) $\approx$ (1771.2 - 763.90) / 0.8641 = 1165.72 hours

Thus, expected remaining life of the radar aged 800 hours is 1165.72 hours.

3.11 MEAN (OPERATING) TIME BETWEEN FAILURE

MTBF stands for *mean operating time between failures* (wrongly mentioned as *mean time between failures* throughout the literature) and is used as a reliability measure for repairable systems. In British Standard (BS 3527) MTBF is defined as follows:

For a stated period in the life of a functional unit, the mean value of the lengths of time between consecutive failures under stated condition.

MTBF is extremely difficult to predict since it depends on several factors such as operating conditions, maintenance and repair effectiveness etc. In fact, it is very rarely predicted with an acceptable accuracy. In 1987 the US Army conducted a survey of the purchase of their SINCGARS radios (example stated in chapter 1) that had been subjected to competitive procurement and delivery from 9 different suppliers. They wanted to establish how the observed Reliability in-service compared to that which had been predicted by each supplier (using MIL-HDBK-217). The output of this exercise is shown in Table 3.8 (Knowles, 1995). It is interesting to note that they are all same radios, same design, same choice of components (but different manufacturers) and the requirement set by the Army was MTBF of 1250 hours with 80% confidence. Majority of the suppliers' observed MTBF was no where near their prediction.

Table 3.8 SINCGARS radios 217 prediction and the observed MTBF

Vendor	MIL-HDBK-217 (hours)	Observed MTBF (hours)
A	7247	1160
B	5765	74
C	3500	624
D	2500	2174
E	2500	51
F	2000	1056
G	1600	3612
H	1400	98
I	1000	472

The reason for huge difference between the predicted and observed MTBF can be attributed to the large number of factors that influence MTBF. Let us assume that the sequence of random variables X_1 , X_2 , X_3 , ...X_n represent the operating time of the item before i-th failure (Figure 3.18). MTBF can be predicted by taking the average of expected values of the random variables X_1 , X_2 , X_3,..., X_n etc. To determine these expected values it is necessary to determine the distribution of the random variables and the corresponding parameters. As soon as an item fails, appropriate maintenance activities will be carried out. This involves replacing the rejected components with either new parts or parts that have been previously recovered (repaired). Each of these components will have a different wear out characteristic governed by a different distribution. To find the expected value of the random variable X_2 one should take into account the fact that not all components of the item are new and, indeed, those, which are not new, may have quite different ages. This makes it almost impossible to determine the distribution of the random variable X_2 and hence the expected value.

The science of failures has not advanced sufficiently, as yet, to be able to predict failure time distribution in all cases. This is currently done empirically by running a sample of items on test until they fail, or for an extended period, usually under 'ideal' conditions that attempt to simulate the operational environment. Military aircraft-engines, for example, are expected to operate while subjected to forces between -5 and + 9 'g', altitudes from zero to 50000 feet (15000 meters) and speeds from zero to Mach 2+. One has to test the equipment with some new and some old components to find the expected values of the random variables X_2, X_3, etc. In practice most of the testing is done on new items with all new components in pristine condition. The value derived by these type of testing will give the expected value of the random variable X_1 . In practice, the expected value of X_1 is quoted as MTBF. In fact, the expected value of X_1 will give only the Mean Time To First Failure (as the testing is done on new items and the times reflect the time to first failure) and not the MTBF. To calculate MTBF one should consider the expected values of the random variables X_2, X_3, etc.

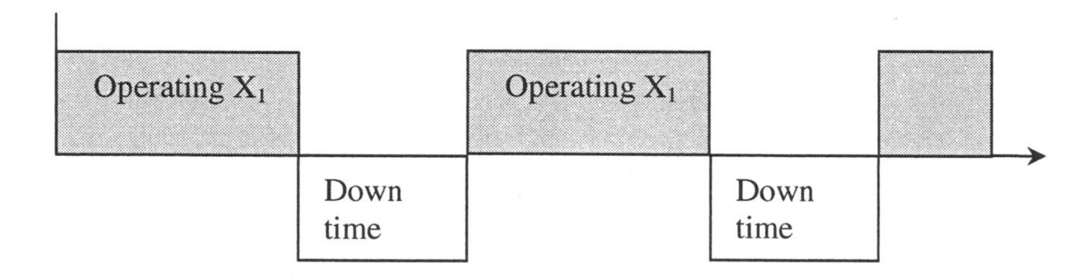

Figure 3.18 operating profile of a generic item

If the time to failure distribution of the system is exponential then the MTBF can be estimated using the following equation (recommended by MIL-HDBK-217 and DEF-STAN-00-40):

$$MTBF = \frac{T}{n} \tag{3.39}$$

where, T is the total operating period and `n' is the number of failures during this period. Note that the above relation is valid only for large value of T. If n = 0, then MTBF becomes infinity, thus one should be careful in using the above relation. The above expression can be used only when sufficient amount of data is available.

Characteristics of MTBF

1. The value of MTBF is equal to MTTF if after each repair the system is recoverd to as good as new.
2. MTBF = $1 / \lambda$ for exponential distribution, where λ is the scale parameter (also the hazard function).

Applications of MTBF

1. For a repairable system, MTBF is the average time in service between failures. Note that, this does not include the time spent at repair facility by the system.
2. MTBF is used to predict steady-state availability measures like inherent and operational availability.

3.12 MAINTENANCE FREE OPERATING PERIOD (MFOP)

Maintenance Free Operating Period is defined as:

The period of operation (for example, for military combat aircraft, a typical MFOP may be 100, 200 or 300 flying hours) during which an item will be able to carry out all its assigned missions, without the operator being restricted in any way due to system faults or limitations, with the minimum of maintenance.

In other words, maintenance free operating period guarantees a certain period of operation without any interruption for unscheduled maintenance. A MFOP (or cycles of MFOP) is usually followed by a maintenance recovery period (MRP). *MRP is defined as the period during which the appropriate scheduled maintenance is carried out.* Since it is almost impossible to give 100 % guaranteed MFOP, we use the concept of *maintenance free operating period survivability* (MFOPS) to measure MFOP. MFOPS is the probability that the part, sub-system or system will survive for the duration of MFOP given that it was in a state of functioning at the start of the period. Note, unlike most warranties, the MFOP will not always apply to new items, indeed, most of the time, the ages of the constituent components will be quite varied and in many cases, unknown. It should be also noted that during MFOP the redundant items are allowed to fail, without causing any unscheduled maintenance.

3.12.1 Maintenance free operating period survivability

Let us consider a system with n components connected in series. If the reliability requirement is MFOP of t_{mf} life units, then the corresponding probability that the system will survive the stated MFOP, given that all the components of the system are new is given by:

$$MFOPS(t_{mf}) = \prod_{k=1}^{n} \frac{R_k(t_{mf})}{R_k(0)} \tag{3.40}$$

where $R_k(t_{mf})$ is the reliability of the k-th component for (the first) t_{mf} life units. The equation (3.40) gives the probability for the system to have MFOP of t_{mf} life units during the first cycle. In general, for i-th cycle (here each cycle refers to each t_{mf} life units), the probability the system will have MFOP of t_{mf} life units is given by:

$$MFOPS(t_{mf}) = \prod_{k=1}^{n} \frac{R_k(i \times t_{mf})}{R_k([i-1] \times t_{mf})} \tag{3.41}$$

MFOP of items with Weibull distributed failure times

For a component with failure mode, which can be modeled by the Weibull distribution the probability of surviving t_{mf} units of time, given that the item has survived t units of time is given by:

$$MFOPS(t_{mf}) = \exp(-\frac{t^{\beta} - (t + t_{mf})^{\beta}}{\eta^{\beta}}) \tag{3.42}$$

where, η is the scale parameter and β is the shape parameter of the Weibull distribution. The MFOP period for a given level of confidence can be calculated by rearranging equation (3.42) as follows:

$$t_{mf} = [t^{\beta} - \eta^{\beta} \ln(MFOPS(t_{mf}))]^{1/\beta} - t \tag{3.43}$$

Maximum length of MFOP

The maximum length of MFOP for a stated MFOPS actually represents the design life of that system. Design life denotes the age of the item up to which the reliability of the system is greater than or equal to the designed reliability value. For example assume that the time-to-failure distribution of the item be Weibull with scale parameter η and shape parameter β. Then the MFOP duration for a specified MFOPS requirement is then given by:

$$MFOP = \eta \times \{\ln\left(\frac{1}{MFOPS}\right)\}^{1/\beta} \tag{3.44}$$

Procedure to calculate the number of cycles the system satisfies the required MFOP.

If the required MFOP is say, t_{mf}, life units and the corresponding MFOPS is α. It may not be necessary to carry out maintenance recovery, after every MFOP. The following steps can be used to find how many such MFOPs can be carried out without any maintenance.

Step 1: Set i = 1.

Step 2: Calculate $MFOP(i,\alpha) = \prod_{k=1}^{n} \frac{R_k(i \times t_{mf})}{R_k([i-1] \times t_{mf})}$

Step 3: If $\prod_{k=1}^{i} MFOP(i,\alpha) \leq \alpha$, then Go To Step 5.

Step 4: i = i + 1, Go To Step 2.

Step 5: Number of cycles is i -1. STOP.

EXAMPLE 3.17

For a computer to be used in a space station, it was required that the MFOP duration for the memory unit should be at least 10000 hours at 95% confidence. It was also required that the memory unit should be screened for 500 hours at different temperature cycles. Two memory chips were available in the market that can be used in the computer to be installed in the space station. The time to failure of a computer memory chip 1 follows Normal distribution with μ = 12000 hours and σ = 1000. The time to failure of a computer memory chip 2 follows Weibull distribution with η = 12000 hours and β = 2.2. Find which chip will satisfy the requirement.

Since the memory unit is subject to temperature screening, the age of the memory unit when put into mission will be 250 hours.

Case 1. Memory chip 1.

MFOPS for the memory chip 1 for the duration of 10000 hours is given by:

$$\text{MFOPS} = \frac{\Phi(\frac{12000-10000}{1000})}{\Phi(\frac{12000-500}{1000})} = \frac{\Phi(2.0)}{\Phi(11.5)} = 0.9772$$

Case 2. Memory chip 2

MFOPS for the memory chip 2 for the duration of 10, 000 hours is given by:

$$MFOPS = \frac{\exp(-(10000/12000)^{2.2})}{\exp(-(1000/12000)^{2.2})} = 0.5141$$

Since MFOPS for chip 1 is greater than 0.95, it will satisfy the MFOP requirement.

3.13 CASE STUDY: ENGINEER TANK ENGINE (ARMOURED VEHICLE ENGINE)

In this section, we discuss a case study on engineer tank (armoured vehicle). To maintain the confidentiality, the names of the user and the manufacturer are not disclosed. Engineer armoured vehicles, also known as Engineer Tank, are used on the battlefield to undertake engineering tasks. This vehicle's primary task on the battlefield is to open routes. Its tasks include providing a route across short gaps, countering mines by clearing routes across mine fields and obstacle breaching and clearance (by digging and bulldozing). The functional elements for the armoured vehicle are: mobility, surveillance, communications, Lethality and survivability. During the development of the engineer tank engine, reliability targets were set and these targets are shown in Table 3.9

Table 3-9. Target reliability for the engineer tank engine

Factor	Engine hours
Mean Time Between Failure (MTBF) [Failures which result in unscheduled maintenance]	420
Mean Time Between Mission Critical Failures (MTBMCF). [Failure which cause loss of propulsion.]	3500
Mean Time Between Routine Overhaul of an engine.	13500

The failure data collected from 70 engines are shown in Table 3.10. These failures have resulted in unscheduled maintenance of the engine.

Table 3-10. In-service failure data of tank engine

215	114	247	122	91	291	194	315	241	9
95	378	425	252	22	153	195	165	130	451
456	350	105	275	294	232	441	94	360	202
18	50	68	244	126	557	168	64	99	323
42	560	232	198	27	947	239	89	325	273
178	33	997	75	126	23	416	247	292	216
210	36	46	223	284	52	223	454	151	88

The time to failure data given in Table 3.10 is analyzed using a statistical software package MINITAB to find the best fit probability distribution. MINITAB uses regression techniques to find the best fit distribution (this will be discussed in Chapter 7). The reliability function associated with the data is shown in Figure 3.19. The best fit in this case is a Weibull distribution with scale parameter η = 246.361 and shape parameter β = 1.286.

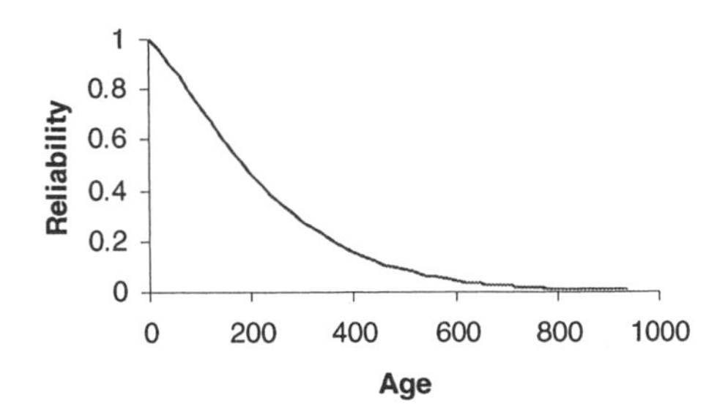

Figure 3-18. Tank engine reliability function

The mean time between failures (which resulted in unscheduled maintenance) is 228.052. Since the target MTBF is 420 engine hours, the corresponding Sigma level is given by:

$$\text{NORMINV}\left(1 - \left(\frac{1}{228.052} - \frac{1}{420}\right), 1.5, 1\right) = 4.3774$$

In this case the process sigma (which developed the engineer tank engine) is 4.3774. **One should note that, not all failures are manufacturing process related. Many failures can be due to the processes used by the user, and hence, failures due to users negligence, misuse etc should be removed from the failure data before assessing sigma level o the manufacturer of the product.**

Chapter 4

SYSTEM RELIABILITY

The best tank terrain is the one without anti-tank weapons
-Russian Military Doctrine

4.1 INTRODUCTION

A system is a collection of components, subsystems or assemblies arranged in a specific pattern in order to achieve desired function(s) with acceptable performance and reliability. The types of components, their quantities, their quality and the manner in which they are arranged within the system have a direct bearing on the reliability of the system. In order to calculate the reliability of a system the reliabilities of the subsystems are calculated and combined using reliability models that are based on certain probability laws. The level to which the decomposition is done must be such that the reliabilities of the resulting components are known with reasonable and acceptable precision. It may not therefore be necessary to decompose the system into individual real components but into a set of devices or subsystems, the reliabilities of which are known from experience. This approach requires knowledge of the physical structure of the system and the nature of its functions sufficiently to determine the behaviour of the system in the event of failure of a subsystem. The system reliability models discussed in this chapter are based on the assumption that the components fail independent of one another; i.e. the failure of one component does not affect the functioning of other components.

In order to understand the relationship between the system and its components for calculating the reliability, the reliability block diagram is constructed. In practice a system is frequently represented as a network in which its components are connected together either in series, parallel or a combination thereof. Here one talks of series and parallel connection in the reliability sense, which can be different from their connection in an electronic circuit, for example. Based on the structure of the reliability block diagram, system reliability measures are derived.

4.2 RELIABILITY PREDICTION

Reliability predictions are most useful during the early phase of system design, before hardware is constructed and tested. It can provide a rational basis for design decisions, choice between alternative concepts, variations in part quality levels, appropriate application of de-rating factors and use of proven versus state-of-the-art methods and other related factors.

Reliability prediction of a system at the inception of its design is also highly desirable since it helps in forecasting the support costs, spares requirements and warranty costs. An accurate prediction implies a thorough knowledge of the causes of failure and the contribution to reliability of individual components. For systems for which the majority, if not all, of the failures are caused by external factors and hence are independent of the ages of the components, the failure rates of the components form the basis for the reliability prediction of the system. Failure rates of the commonly used standard electronic and mechanical components have been listed in certain handbooks or in the literature. These failure rates are based on historical data collected on the reliability of similar components from various sources. The constant failure rates listed in the literature are obtained assuming the exponential distribution for failure times. Great care should be taken while using these as they may not take into account, the peculiar circumstances relevant to the system under study. The following are some of the published documents which are used to obtain the failure rates of the components:

1. MIL-HDBK-217F (Notice 2) - for electronic components
2. Non-electronic Parts Reliability Data (NPRD -95) – for mechanical components
3. BellCore / TelCordia (TR-332) for telecommunication components

The failure rates are typically expressed as failures per million hours and the above handbooks have used the same unit for the failure rates. [The unit hour is used in a generic sense and it can have units like miles/kilometers in case of a vehicle, cycles in case of a bread toaster, number of landings in case of aircraft wheels].

The failure data of the components can also be obtained from other sources, including in-house reliability tests, accelerated life tests, field data, warranty data, engineering knowledge and similarity to prior design. When possible, it is strongly advised to use data obtained in this way in preference to any published data.

4.2.1 Duty cycle

In a system all the components may not be active during the total operational time. In some cases, if the actual mission period is t hours, some components of the system may have to operate more than t hours (in many cases it can be less than t hours). An aircraft jet engine will be switched on at least 10 minutes before the actual flight although this time is frequently ignored when estimating time-to-failure distribution parameters. Thus for 10 hours flight, the engine may have to operate more than 10 hours. In an automobile the headlights are usually on only during night driving. The brake is applied only when required and the clutch is operated only when changing gear. Thus only when a component operates continuously during systems operation, is the component's operating time is equal to the system's operating time. If the component operates for t_1 hours in t system operating hours, it assumes in the system's time scale a failure rate:

$$\lambda = \lambda'\left(\frac{t_1}{t}\right) \tag{4.1}$$

Where λ' is the failure rate of the component while in operation. The factor (t_1/t) is referred to as the duty cycle and in general is expressed as a percentage.

4.2.2 Cyclic exchange rate

In case of aircraft jet engines, for example, although the time it is operational may be longer than the flight time, what is often more important from the viewpoint of reliability is the way in which the engines are actually used. For many components, the primary cause of failure is the amount of stress experienced rather than the number of hours flown. Estimates of this can vary from the assumption that each flight is equivalent to one stress cycle. This is due to the advent of highly sophisticated on-board engine monitoring systems which can take many thousands of measurements of temperatures, pressures and spool speeds during each flight and convert these through special algorithms into the number of stress cycles. Typically, for commercial long-haul aircraft, the one-cycle-per-stage-length estimate is perfectly adequate, but for fast combat military aircraft, it requires the much more sophisticated approach. In cases where not all aircraft in a fleet [of military aircraft] are fitted with engine monitoring equipment, it may be necessary to use what is generally referred to as a cyclic-exchange rate to convert flying hours into stress cycles.

The following steps are involved in prediction of the reliability of a system:

1. Construction of the Reliability Block Diagram (RBD) of the system. This may involve performing Failure Modes and Effects Analysis (FMEA).
2. Determination of the operational profile of each block in the diagram.
3. Derivation of the time to failure distribution of each block.
4. Derivation of the Life Exchange Rate Matrix (LERM) for the different components within the system.
5. Computation of the reliability functions of each block.
6. Computation of the reliability function of the system

4.3 RELIABILITY BLOCK DIAGRAM

The Reliability Block Diagram (RBD) of a system is a logical diagram of several blocks connected in series, parallel, standby or combinations thereof as appropriate, which results in the successful performance of the system. Each block in the diagram represents a component or subsystem, which is required for the functionality of the system. The connecting lines between the blocks do not necessarily represent any physical connections in the system. For example, the bus architecture in a system may be represented by a single block in the RBD. A reliability block diagram does not necessarily represent the systems operational logic or functional partitioning of the system. The structure of the RBD is determined by the effect of failure of each block on the successful functionality of the system as a whole. The blocks whose failure causes the failure of the system are connected in series and the blocks whose failure alone will not cause the failure of the system are connected in parallel. The system architecture and design documents can be used for this purpose.

The first level of RBD will consist of major subsystems required for the successful performance of the system. It is possible that each block of the RBD representing a subsystem is further broken down into more blocks depending on the information available for the particular block. This is illustrated by an example (Figure 4.1) RBD drawn for an automobile.

Figure 4.1 shows the minimum subsystems required for running an automobile. The blocks are connected in series implying that every one of the blocks is essential for the successful running of the automobile. Figure

4.2 shows the subsystems of the engine system. We see a redundancy for starting the engine. We may use the starter, jump-start or manually crank the engine. "Jump start" means using another battery – usually in another automobile. Although manual cranking is almost obsolete today, there are some special types of vehicles, e.g. tractors, which can still be started this way.

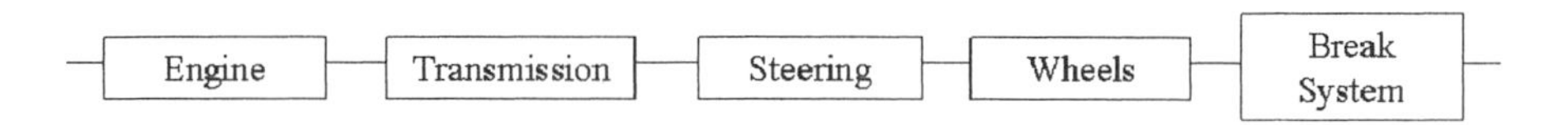

Figure 4-1. Reliability block diagram of an automobile

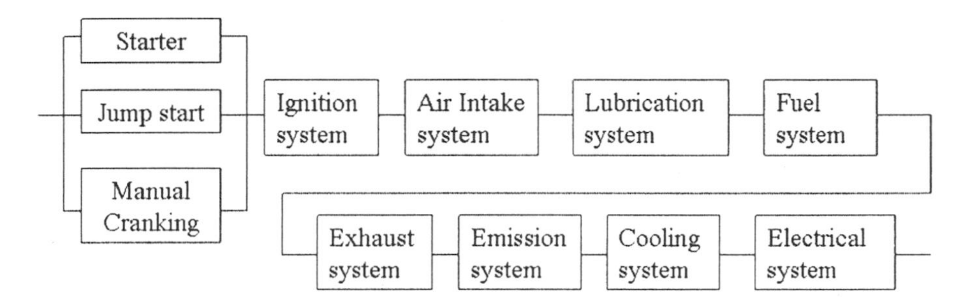

Figure 4-2. Reliability block diagram of an engine

4.4 SERIES SYSTEM

Series systems are those in which every one of the components is required to be in a state of functioning (SoFu) for the system to be in a SoFu. Figure 4.1 is an example of a series system. To derive the expression for the reliability of a series system, consider three components C1, C2, and C3 connected in series as shown in Figure 4.3.

Let $R_s(t)$ be the probability that the system is in a SoFu after time **t** hours from new and $R_i(t)$ be the corresponding probabilities for the components (i = 1…3). The requirement for the system success is that all the components must be working. Thus if all the failures are *independent*, then:

$$R_s(t) = R_1(t) \cdot R_2(t) \cdot R_3(t) \tag{4.2}$$

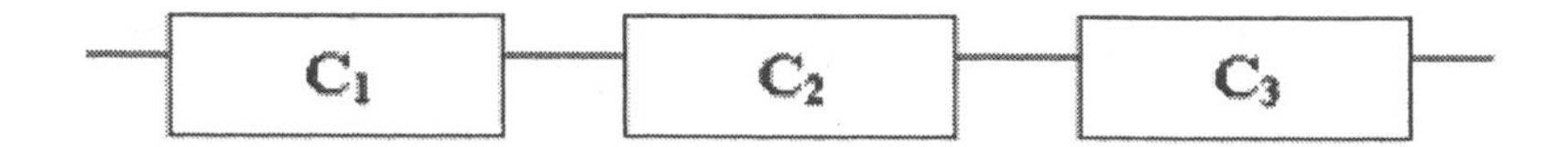

Figure 4-3. Series system

By *independent*, we mean that the failure of a component in the system does not affect the likelihood of failure of any other component in the system. If there are ***n*** components in series in a system then the reliability of the system is given by:

$$R_s(t)=\prod_{i=1}^{n} R_i(t) \tag{4.3}$$

Equation (4.3) is referred to as the product rule of reliability since it establishes that the reliability of a series system is the product of the individual component reliabilities. The unreliability (Failure) and reliability of a system are complimentary to each other. Thus the unreliability, $F_s(t)$, of a series system with ***n*** components is given by:

$$F_s(t)=1-\prod_{i=1}^{n} R_i(t) \tag{4.4}$$

Each block in the RBD can have a different failure distribution. The reliability of each block is obtained and the product of all the reliabilities is taken to arrive at the system reliability. If the components follow the exponential distribution then the reliability of the system reduces to:

$$R_s(t)=e^{-\lambda_1 t}\cdot e^{-\lambda_2 t}\cdot\ldots\cdot e^{-\lambda_n t}=e^{-\left(\sum_{i=1}^{n}\lambda_i t\right)} \tag{4.5}$$

The failure rate of the system can easily be derived from equation (4.5), and is given by:

$$\lambda_s=\sum_{i=1}^{n}\lambda_i \tag{4.6}$$

In general most of electronic components are assumed to follow the exponential distribution for their failure times and mechanical components follow the Weibull or log-normal distribution. In summary the system reliability is the product of the component reliabilities and the system failure rate is the sum of the component failure rates for a series system in which all the times to failure are exponentially distributed.

EXAMPLE 4.1

Consider a computer system consisting CPU, monitor and printer, which are in series for its functionality. The reliabilities of these subsystems (components) for a given period t are: $R_1(t) = 0.7$, $R_2(t) = 0.9$, $R_3(t) = 0.8$ for the CPU, monitor and printer respectively. Find the reliability of the system.

Using equation (4.2) we find that the reliability $R_s(t)$ of the computer system is given by

$$R_S(t) = R_1(t) . R_2(t) . R_3(t) = (0.7)(0.9)(0.8) = 0.504$$

The reliability of a series system is always less than or equal to the minimum of the reliabilities of the individual components in it. This is seen from the example 4.1. Thus for a given time t the system reliability $R_s(t)$ of a series system is:

$$R_s(t) \leq Min(R_1(t), R_2(t) \dots R_n(t)) \tag{4.7}$$

Reliability prediction serves as a useful tool in the development of alternatives to a particular design. For example, if a reliability goal is set as $R_G = 0.85$ in example 4.1 then the question arises as to how this can be achieved, if indeed it is even possible. Let us examine this by increasing the reliability of each of the subsystems/components one at a time and look at the effect on the system reliability. Figure 4.4 shows the graph of system reliability versus individual component reliability. For example the system reliability is calculated by increasing the reliability of C1 from 0.7 to 0.99 in steps keeping the reliability of the other two components same. Similar exercise is repeated for the other two components also and three curves are shown in the figure 4.4, one for each component. Figure 4.4 shows that by raising the reliability of a single component (C1, C2 or C3) the reliability goal 0.85 cannot be achieved. The next trial can be increasing the reliability of two components at a time. Then the question arises as to which two components have to be considered for the reliability improvement. [In this example, one can verify, if all the components reach a value of 0.948 then the reliability goal 0.85 will be achieved]. In reality it is not easy to increase the reliability for any component with such high increments. In some cases it will be possible to increase the reliability of a component through the provision of parallel component(s). If a manufacturer is given target reliability at the system level, to achieve higher sigma level, they may have to use redundancy at component level.

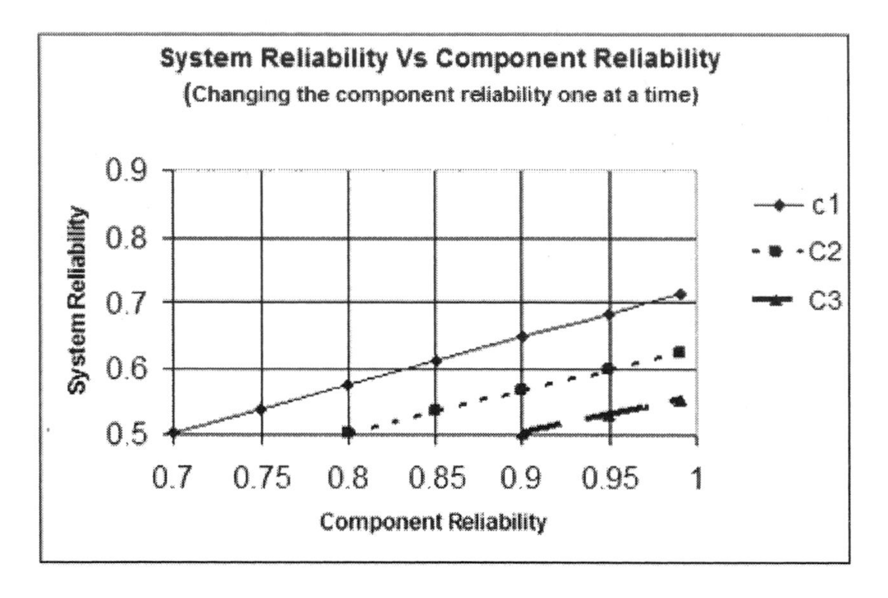

Figure 4-4. Change in system reliability due to increase in component reliability

EXAMPLE 4.2

Consider a system having four components (C1, C2, C3, C4). The time to failure distributions of the four components are as follows:

C1	Weibull distribution	$\eta = 1000$ hours, $\beta = 3.2$
C2	Exponential distribution	$\lambda = 0.002$ / hour
C3	Normal distribution	$\mu = 400$ hours, $\sigma = 50$ hours
C4	Weibull distribution	$\eta = 300$ hours, $\beta = 2.4$

Find the reliability of the system for the first 100 hours.

Reliability of component 1

$$R_1(100) = e^{-\left(\frac{t}{\eta}\right)^{\beta}} = e^{-\left(\frac{100}{1000}\right)^{3..2}} = 0.9994$$

Reliability of component 2

$$R_2(100) = e^{-\lambda t} = e^{-(0.002\times 100)} = 0.8187$$

Reliability of component 3

$$R_3(100) = \Phi\left(\frac{\mu - t}{\sigma}\right) = \Phi\left(\frac{400 - 100}{50}\right) = 1.0000$$

Reliability of component 4

$$R_4(100) = e^{-\left(\frac{t}{\eta}\right)^{\beta}} = e^{-\left(\frac{100}{300}\right)^{2.4}} = 0.9309$$

Reliability of the system

$R_s(100) = R_1(100)$ x $R_2(100)$ x $R_3(100)$ x $R_4(100) = 0.7617$

4.4.1 Mean time to failure of a series configuration

The mean time to failure of any system is given by

$$MTTF = \int_0^\infty t\, f(t)\, dt = \int_0^\infty R_s(t) dt \tag{4.8}$$

where f(t) is the probability density function of the failure times. The mean time to failure of a series configuration denoted by $MTTF_s$ is given by

$$MTTF_s = \int_0^\infty R_s(t)dt = \int_0^\infty \prod_{i=1}^{n} R_i(t)dt \tag{4.9}$$

Numerical methods can be used to evaluate the above integral. If the time to failure distribution is exponential then the $MTTF_s$ of the series system is given by

$$MTTF_s = \int_0^\infty \prod_{i=1}^{n} R_i(t)dt = \int_0^\infty \prod_{i=1}^{n} e^{-\lambda_i(t)}\,dt = \int_0^\infty e^{-\sum_{i=1}^{n}\lambda_i t}\,dt = \frac{1}{\sum_{i=1}^{n}\lambda_i} \tag{4.10}$$

Thus the $MTTF_s$ of a series system is given by the reciprocal of the sum of the failure rates of the components in the system. This result is true only when the time to failure distribution of all components is exponential. Even if one of the components follows any distribution other than exponential, the result in equation (4.10) is not valid.

4.4.2 Sigma level and $MTTF_s$

Let $MTTF_T$ and $MTTF_A$ denote the target and achieved mean time to failure of a series configuration respectively. The DPU for not meeting the target system reliability, measured in terms of MTTF, is given by:

$$DPU = \left[\frac{1}{MTTF_A} - \frac{1}{MTTF_T}\right], \quad MTTF_A < MTTF_T \tag{4.11}$$

Equation (4.11) gives the additional number of system failures one can expect per unit time for not meeting the target MTTF. The corresponding DPMO is given by:

$$DPMO = [\frac{1}{MTTF_A} - \frac{1}{MTTF_T}] \times 10^6 \tag{4.12}$$

Equation (4.12) gives the additional number of failures one can expect if the system is used for one million life units due to not meeting the target

MTTF. The sigma level can be calculated using equation (3.9), where the value of P is given by:

$$P = \sqrt{\ln\left(\frac{1}{\left(\frac{1}{MTTF_A} - \frac{1}{MTTF_T}\right)^2}\right)} \tag{4.13}$$

The corresponding Excel® function is given by:

$$\text{NORMINV}\left(1 - \left(\frac{1}{\text{MTTF}_\text{A}} - \frac{1}{\text{MTTF}_\text{T}}\right), 1.5, 1\right) \tag{4.14}$$

EXAMPLE: 4.3

An electronic circuit to drive a special double-coil relay consists of 4 (metal film) resistors, two general-purpose diodes and two pairs of Darlington transistors. Find the failure rate, reliability for a continuous operation of 100 hours and the MTTF for the circuit.

Table 4-1. Failure rate of different components

Component	Unit failure rate Failures/10^6 hours	Quantity	Total failure rate Failures/10^6 hours	% contribution of failure rate
Resistors	0.040834	4	0.163335	39 %
Diodes	0.010425	2	0.020850	5 %
Transistors	0.051397	4	0.205588	49 %
Interconnection	0.031200	1	0.031200	7 %
Total Failure rate of the circuit			0.420974	
Reliability for t = 100 hours = 0.99996				
MTTF = 3.9 x 10^6 hours				

The failure rates are taken from the MIL-HDBK-217F for ground benign environment. Table 4.1 gives the unit failure rate and the total failure rate for the components in the circuit. The total failure rate, reliability and MTTF for the circuit are obtained using equations (4.6), (4.5) and (4.10) respectively.

In the table 4.1 apart from the mentioned components in the example the interconnection is also taken into account. This takes into account the failure due to the type of soldering or mounting technology of components on the printed circuit board, which often plays a major role in circuit failure. The table also gives the percentage contribution of the component failure rates in the circuit. Tabulation of this information above is very useful as it enables one to determine the relative contribution of the different components to failure of the circuit. Based on the importance, criticality and duty cycle of the component the reliability of the circuit can be improved by using a better component with lower failure rate or providing spares for such components. The calculated MTTF shows an average period of time to fail. One should not mistake it for a guaranteed failure free operation period.

EXAMPLE 4.4

Suppose we know that a particular circuit has an MTTF of 1000 hours. What is the probability of getting zero failures in a total of 1000 hours of operation?

$$\text{Failure rate} = \lambda = \frac{1}{MTTF} = \frac{1}{1000} = 0.001$$

Time of operation t =1000 hours and λt =1000 * 0.001 = 1.0

Using Poisson distribution, the probability of zero failures, P(0) is given by:

$$P(0) = \frac{e^{-\lambda t}(\lambda t)^0}{0!} = e^{-1} = \frac{1}{e} = 0.37(or\ 37\%)$$

This shows that the chance of having zero failures is about 37%. **This important result states that there is only approximately one in three chances of surviving to the MTTF without a failure!**

4.4.3 Hazard rate (function) of a series system

The hazard function h(t) or instantaneous failure rate is given by

$$h(t)=\frac{f(t)}{R(t)}=\frac{F'(t)}{R(t)}=\frac{-R'(t)}{R(t)} \tag{4.15}$$

Where $f(t)$ is the probability density function and $F(t)$ is the cumulative density function

$$F(T)=\int_0^T f(t)dt$$

For an exponential time to failure distribution

$$f(t)=\lambda e^{-\lambda t} \quad ,F(t)=1-e^{-\lambda t}$$
$$R(t)=e^{-\lambda t} \quad \textit{and} \quad R'(t)=-\lambda e^{-\lambda t} \text{ and } h(t)=\lambda$$

Consider a system with n components connected in series. Let λ_1, λ_2, λ_3, ... λ_n be the failure rates of the n components. The reliability R_s of the system is given by equation (4.5)

$$R_s = e^{-\sum_{i=1}^{n} \lambda_i t}$$

$$h(t) = \frac{\sum \lambda_i \; e^{-\sum \lambda_i t}}{e^{-\sum \lambda_i t}} = \sum \lambda_i \tag{4.16}$$

In particular if the systems have equal failure rates, say $\lambda_1 = \lambda_2 = \ldots = \lambda_n = \lambda$ then h(t) = nλ. Thus we see that the failure rate is a constant and is independent of time in this very special case when all the failure distributions are exponential.

This is an important result which shows that if the time to failure distribution of a system is exponential then the hazard rate is constant and hence independent of time. This also establishes the **lack of memory property** of the exponential distribution. The probability that the component fails in the next hour of operation is the same, irrespective of the time duration for which it has been operating. In a probabilistic sense the component does not age, degrade or wear out so the failure has a pure random chance of happening.

4.5 LIFE EXCHANGE RATE MATRIX

Not all the components of the system will have the same utilization of life unit. The operational environment can also change the ageing pattern of different components within a system. For example if an automobile is run for t hours, clutch, brake and starter are operated in several cycles depending on the driving conditions of the road. The life of the wheels is largely dependent on the distance covered (miles). The life of the automobile engine is dependent on the hours it is run whereas the life of the clutch is related to the number of gear changes, the brakes to the number of times they are used and the starter to the number of journeys. Another example is the average flight of a domestic flight within Japan is around 30 minutes compared to that of India, which is around 90 minutes. Thus for a given number of flying hours there will be many more operations of the landing gear in Japan than in India. Thus it is clear that flying hours alone cannot be used to determine the life of the landing gear, a subsystem of the aircraft. It is very common that different items within a system may have different life units such as hours, miles, cycles, flying hours, landing cycles etc. Thus, to find the reliability of a system whose items have different life units, it is necessary to normalize the life units. In this section we introduce the concept of life exchange rate matrix, which can be used to describe the exchange rates between various life units.

Life exchange rate matrix (LERM) is a square matrix of size n, where n is the number of items in the system. Let us denote the life exchange rate matrix as $R = [r_{i,j}]$ where $r_{i,j}$ is the $(i,j)^{th}$ element in the LERM. Thus for a system with n items connected in series, the LERM can be represented as [U Dinesh Kumar *et al* 2000]

$$LERM = \begin{bmatrix} r_{1,1} & r_{1,2} & \cdots & r_{1,n} \\ r_{2,1} & r_{2,2} & \cdots & r_{2,n} \\ \vdots & \vdots & \vdots & \vdots \\ r_{n,1} & r_{n,2} & \cdots & r_{n,n} \end{bmatrix} \quad (4.17)$$

The elements of LERM are interpreted as follows:

$r_{i,j}$ denotes that :

$$1\, life\ unit\ of\ i = r_{ij} \times 1\, life\ unit\ of\ j$$

Any LERM will satisfy the following conditions:

$$r_{i,i} = 1 \quad \forall\, i,$$
$$r_{i,j} = r_{i,k} \times r_{k,j} \quad \forall i, j, k \quad \text{where}$$
$$r_{i,j} = \frac{1}{r_{j,i}}$$

As an example, let us consider a system with three items connected in series configuration (Figure 4.5). Let the life unit of items 1, 2 and 3 be hours, miles and cycles respectively.

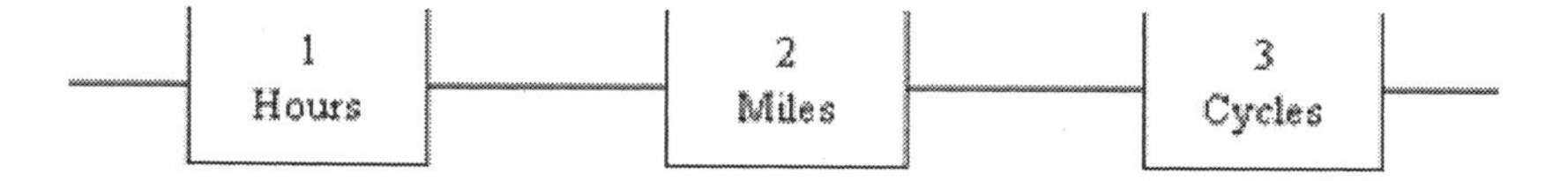

Figure 4-5. Series system with different life exchange rates

Assume that:

1 hour = 10 miles
1 hour = 5 cycles

Using the above data it is easy to construct the life exchange rate matrix for the above system. The LERM for the above example is

$$R = \begin{bmatrix} 1 & 10 & 5 \\ 1/10 & 1 & 1/2 \\ 1/5 & 2 & 1 \end{bmatrix} = \begin{bmatrix} 1 & 10 & 5 \\ 0.1 & 1 & 0.5 \\ 0.2 & 2 & 1 \end{bmatrix}$$

One can verify that the above matrix satisfies all three conditions for a life exchange rate matrix. Using the above matrix, one can easily measure reliability characteristics in normalized life unit. For the RBD shown in

Figure 4.5, reliability of the system for 5 cycles is given by R_1 (1) x R_2 (10) x R_3 (5).

EXAMPLE 4.5

Reliability block diagram of a system consists of three modules A, B and C connected in series. The time to failure of module A follows a Weibull distribution with scale parameter η = 100 hours and β = 3.2. The time to failure of module B follows the normal distribution with mean μ = 400 cycles and standard deviation σ = 32 cycles. The time to failure of module C follows the exponential distribution with parameter λ = 0.00015 per mile. It was also noted that during 1 hour, module B performs 12 cycles and module C performs 72 miles. Find the probability that the system will survive up to 240 cycles of module B.

For the system to survive 240 cycles, module A will need to survive up to 20 hours and module C, 1440 miles. The reliabilities of the individual modules are given by :

$$R_A(t_A) = e^{-\left(\frac{t_A}{\eta}\right)^{\beta}} = e^{-\left(\frac{20}{100}\right)^{3.2}} = 0.9942$$

$$R_B(t_B) = \Phi\left(\frac{\mu - t_B}{\sigma}\right) = \Phi\left(\frac{400 - 240}{32}\right) = 1$$

$$R_C(t_C) = e^{-\lambda t_C} = e^{-(0.00015 \times 1440)} = 0.8057$$

The system reliability for 240 cycles is given by:

$$R_s(240) = R_A(20) \times R_B(240) \times R_C(1440) = 0.8010$$

4.6 CONDITIONAL PROBABILITIES OF SURVIVAL

Suppose the system in example 4.5 had already achieved 25 hours of operation without any of the three components failing. What is the probability now that it will survive a further 240 cycles?

To solve this problem, we need to consider conditional probabilities. This is the probability that a module will not fail in the next t units given that it has survived (or is aged) T units. This is not the same as the probability that the module will survive t + T units because this does not take into account the fact that we know it has already survived T units during which time there was a probability of F(T) that it would have failed. If we consider the human population, the life expectancy of a female born in the UK is about 75 years. This means that a new-born baby will on average live to 75 but we know some will die from childhood illnesses (infant mortality), some will die from accidents (e.g. killed on the road, in a train or plane crash, hit by lightning, drown in floods or be murdered) or from adult illnesses such as cancer or heart attacks. A lady aged 50 therefore will have a higher life expectancy but certainly not as high as 125 (which is what it would be if the reliability distribution was exponential). For a lady aged 75 today, we could expect her to live maybe another 15 years on average. Of course some will die before their 76th birthday whilst others may become centenarians, but for our lady to have reached 75 she must have been one of the lucky 30-40%.

The probability that an item will survive a further time t units, given it has already survived T units is

$$R(t+T\mid T)=\frac{R(t+T)}{R(T)} \tag{4.18}$$

Note, for a new item, T = 0 and R(0) = 1 giving R(t+0)/R(0) = R(t). Returning to example 4.5, the conditional probabilities are given by:

$$R_A(t_A+T_A\mid T_A)=\frac{e^{-\left(\frac{(t_A+T_A)}{\eta}\right)^{\beta}}}{e^{-\left(\frac{T_A}{\eta}\right)^{\beta}}}=\frac{e^{-\left(\frac{45}{100}\right)^{3.2}}}{e^{-\left(\frac{25}{100}\right)^{3.2}}}=\frac{0.9253}{0.9882}=0.9363$$

$$R_B(t_B+T_B\mid T_B)=\frac{\Phi\left(\dfrac{\mu-t_B-T_B}{\sigma}\right)}{\Phi\left(\dfrac{\mu-T_B}{\sigma}\right)}=\frac{\Phi\left(\dfrac{400-540}{32}\right)}{\Phi\left(\dfrac{400-300}{32}\right)}=6.0818\times10^{-6}$$

$$R_C(t_C+T_C\mid T_C)=\frac{e^{-\lambda(t_C+T_C)}}{e^{-\lambda(T_C)}}=\frac{e^{-(0.00015\times3240)}}{e^{-(0.00015\times1800)}}=0.8057$$

The probability that the system will survive another 20 hours (240 cycles or 1440 miles) is

$$R_s(20+25\,|\,25) = R_A(45\,|\,25) \times R_B(540\,|\,300) \times R_C(3240\,|\,1800)$$
$$= 4.59 \times 10^{-6}$$

Note: the probability that the system would have survived 45 hours from new is 3.46×10^{-6}. Note also the conditional probability for Module C is exactly same as before.

4.7 PARALLEL SYSTEMS

Parallel systems are those in which the system failure will occur only when all the components in the system fail. Let us consider a system with two components C1 and C2 connected in parallel, reliability-wise, as shown in Figure 4.6. Let R1 and R2 be the reliabilities of the two components for a given time period t. Assuming that the component failures are independent; the system reliability can be obtained as follows:

$$\begin{aligned} R_s(t) &= 1 - F_s(t) = 1 - (1 - R_1(t) \cdot (1 - R_2(t))) \\ &= R_1(t) + R_2(t) - R_1(t) \cdot R_2(t) \end{aligned} \tag{4.19}$$

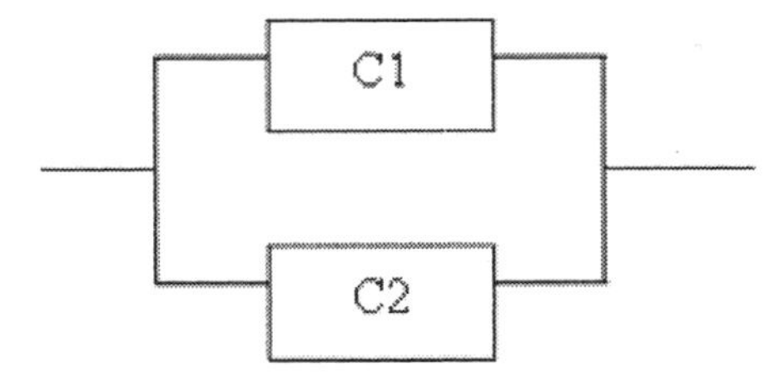

Figure 4-6. Parallel configuration

Suppose the system has ***n*** components connected in parallel then the system reliability function and failure function are given by

$$R_s(t) = 1 - \prod_{i=1}^{n} (1 - R_i(t)) \tag{4.20}$$

$$F_s(t) = \prod_{i=1}^{n} (1 - R_i(t)) \tag{4.21}$$

Equation (4.21) is referred to as the product rule of unreliability, since it establishes that the unreliability of a parallel system is the product of the individual components' unreliability. In particular if a system has n components in parallel with time to failure distribution as exponential with failure rates as λ_1, λ_2, λ_3, ... λ_n, then the reliability of the system is given by

$$R_s(t) = 1 - \prod_{i=1}^{n} \left(1 - e^{-\lambda_i t}\right) \tag{4.22}$$

In particular if $\lambda_1 = \lambda_2 = \ldots \lambda_n = \lambda$ then equation (4.22) reduces to

$$R_s(t) = 1 - \left(1 - e^{-\lambda t}\right)^n \tag{4.23}$$

EXAMPLE 4.6

A system has two components with reliability 0.7 and 0.8 for a given period of time t connected in parallel. Find the reliability of the system.

Applying Equation (4.20) we get,

$R_s(t) = 1-(1-0.7)(1-0.8) = 0.94$

In a parallel system the reliability of the system is always greater than the maximum value of the reliability of the components in the system which is also seen in this example. Thus we have for a system with n components connected in parallel

$$R_s(t) \geq Max\{R_1(t), R_2(t) \ldots R_n(t)\} \tag{4.24}$$

EXAMPLE 4.7

Component of a system has a reliability of 0.7 for a period of time t. Find how many such components should be connected in parallel in order to achieve a reliability goal of 0.95

Using the equation (4.20) we have

$$0.95 = 1-(1-0.7)^n$$

Solving for n from the above equation we get n=2.5. Since n has to be an integer we require 3 components to be connected in parallel to achieve the reliability goal of at least 0.95.

EXAMPLE 4.8

Calculate the reliability of a parallel configuration as **n** (no. of components) increases, for various component reliability values say R=0.7, 0.8, 0.9. Evaluate the equation (4.19) for R=0.7, 0.8 and 0.9 for n = 1,2,3,...,7.

It can be seen from the table 4.2 that for each value of the component reliability, every component added in parallel brings down the failure rate by the same factor. As is to be expected, this factor increases with increase in the individual component reliability.

A look at the curve in Figure 4.7 may suggest that the incremental improvement in reliability with each additional component decreases monotonically with increasing n (no. of components). This can be seen to be illusory if one remembers that a reliability of 1.0 is absolute perfection. The curve approaches the value 1.0 asymptotically.

From the table 4.2 we see that to achieve a reliability of 0.99 we require 4 components with reliability 0.7 each, 3 components of reliability 0.8 or two components of reliability 0.9. Thus it is clear that the number of components to be connected in parallel to achieve a certain reliability goal is dependent on the individual component reliability. While increasing the number of components to achieve a set goal of reliability, the designer must bear this in mind vis-à-vis the cost (money, weight, power consumption and space) that increase linearly with the number of components. Thus the decision to increase the redundancy should not be taken arbitrarily.

Table 4-2. Increase in system reliability of a parallel system as the number of components increase

Component reliability	n	R	Failure Probability (F_n = 1-R_n)	Failure probability Reduction factor (F_n/F_{n-1})
0.7	1	0.7	0.3	Not applicable
	2	0.91	0.09	3.33
	3	0.973	0.027	3.33
	4	0.9919	0.0081	3.33
	5	0.99757	0.00243	3.33
	6	0.99927	0.000729	3.33
	7	0.99978	0.000219	3.33
0.8	1	0.8	0.2	Not applicable
	2	0.96	0.04	5
	3	0.992	0.008	5
	4	0.9984	0.0016	5
	5	0.99968	0.00032	5
	6	0.999936	6.4E-05	5
	7	0.999987	1.28E-05	5
0.9	1	0.9	0.1	Not applicable
	2	0.99	0.01	10
	3	0.999	0.001	10
	4	0.9999	1E-04	10
	5	0.99999	1E-05	10
	6	0.999999	1E-06	10
	7	1.0	1E-07	10

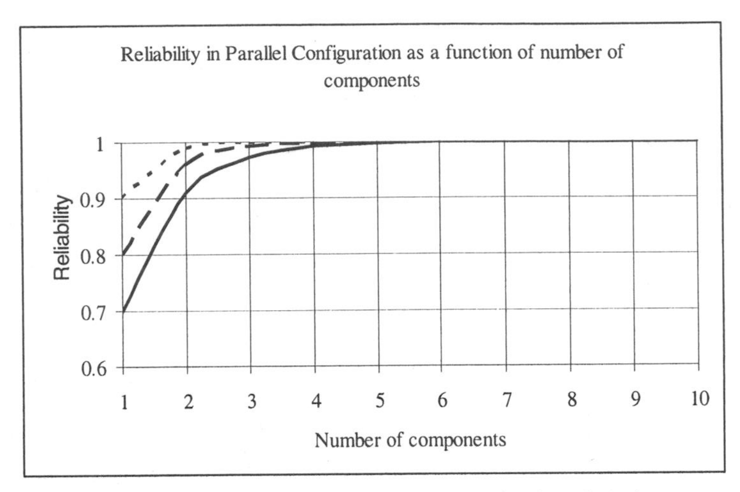

Figure 4-7. Reliability of a parallel system as a function of number of components

4.7.1 Mean time to failure of a parallel configuration

The mean time to failure of a parallel configuration, denoted by $MTTF_s$ is given by:

$$MTTF_s = \int_0^\infty R_s(t)dt = \int_0^\infty \left[1 - \prod_{i=1}^{n}(1 - R_i(t))\right]dt = \int_0^\infty 1 - (1 - R_i(t))^n \, dt \quad (4.25)$$

Let the time to failure distribution be exponential with mean (1/λ). Let there be n similar components connected in parallel. Then the reliability R(t) is given by

$$R(t) = 1 - \left(1 - e^{-\lambda t}\right)^n \quad (4.26)$$

The MTTF for the system is given by

$$MTTF_s = \int_0^\infty \left(1 - \left(1 - e^{-\lambda t}\right)^n\right)dt \quad (4.27)$$

For particular values of n we can simplify the above integrals and derive an expression for $MTTF_s$.

Case 1: n = 2

$$MTTF_s = \int_0^{\infty} 1 - \left(1 - e^{-\lambda t}\right)^2 dt$$

$$= \int_0^{\infty} \left[2e^{-\lambda t} - e^{-2\lambda t}\right] dt$$

$$= \frac{2}{\lambda} - \frac{1}{2\lambda} = \frac{3}{2\lambda}$$

The above expression can also be written as

$$\frac{3}{2\lambda} = \frac{1}{\lambda}\left(1 + \frac{1}{2}\right)$$

Case n = 3

$$MTTF_s = \int_0^{\infty} 1 - \left(1 - e^{-\lambda t}\right)^3 dt$$

$$= \int_0^{\infty} \left(3e^{-\lambda t} - 3e^{-2\lambda t} + e^{-3\lambda t}\right) dt$$

$$= \frac{3}{\lambda} - \frac{3}{2\lambda} + \frac{1}{3\lambda} = \frac{11}{6\lambda}$$

The above expression can again be written as

$$\frac{11}{6\lambda} = \frac{1}{\lambda}\left(1 + \frac{1}{2} + \frac{1}{3}\right)$$

It can be easily shown that for n similar components in parallel

$$MTTF_s = \left(\frac{1}{\lambda}\right)\sum_{k=1}^{n}\frac{1}{k} \tag{4.28}$$

For larger values of n the equation can be reduced to (Aggarwal, 1993)

$$MTTF = \left(\frac{1}{\lambda}\right)\left(Ln(n) + 0.577 + \frac{1}{2n}\right) \tag{4.29}$$

It is to be noted that the averaging of TTF is done over the interval t = 0 to infinity on a failure rate which increases with time (see section 7.1). Therefore, the numerical value obtained is a questionable practical value, where one is concerned with finite time scales. The rigorous calculation of MTTF above is more of an academic value.

4.7.2 Hazard function for a parallel configuration

Let n systems be connected in parallel. Using equations (4.15) and (4.20) we get:

$$h(t) = -\frac{\left(n R'(t)(1 - R(t))^{n-1}\right)}{1 - (1 - R(t))^n} \tag{4.30}$$

Now let us consider the case where the n systems have their time to failure distribution as exponential. Let $\lambda_1, \lambda_2, \lambda_3, \ldots \lambda_n$ be the failure rates of the n systems. Reliability $R_s(t)$ for n systems is given by equation (4.22)

$$R_s(t) = 1 - \prod_{i=1}^{n}\left(1 - e^{-\lambda_i t}\right)$$

The hazard function is given by

$$h(t) = \frac{n\lambda\left(e^{-\lambda t}\right)\left(1 - e^{-\lambda t}\right)^{n-1}}{1 - \left(1 - e^{-\lambda t}\right)^n} = n\lambda B_n(t) \tag{4.31}$$

Where $B_n(t)$ is given by:

$$B_n(t) = \frac{(e^{-\lambda t})(1-e^{-\lambda t})^{n-1}}{1-(1-e^{-\lambda t})^n}$$

Equation (4.31) clearly shows that the failure rate of the parallel system with components having constant failure rate is no more a constant quantity but an increasing function of time. It can be shown that as t increases B_n increases monotonically. This has been proved by showing dh(n)/dt > 0 [Grosh, (1989)] . A simple algebraic proof has also been given in Grosh (1982). Thus the parallel system with constant failure rate has increasing failure rate. Also it can be shown mathematically that:

$$B_n \rightarrow \frac{1}{n} \quad as \quad t \quad \rightarrow \infty$$

For $t \rightarrow \infty$, h(t) tends to λ, which shows that the parallel system can be expected to exhibit a constant failure rate equal to that of a single component, when **t** tends to infinity.

4.8 SERIES-PARALLEL COMBINATION CONFIGURATION

For systems with series-parallel configurations we need to apply the product rule of reliability and unreliability repeatedly to get the reliability of the system. Figure 4.8 shows a typical configuration of a series parallel combination. Let R1, R2, R3, R4, R5 and R6 be the reliabilities of the components C1, C2...C6. The way we go about calculating the reliability of the whole system shown in Figure 4.8 is as follows:

Step 1: Calculate the reliability of C1 & C2 in series – say R' = R1 × R2
Step 2: Calculate the reliability of C3 & C4 in parallel – say R'' = (R3+R4-R3R4)
Step 3: Calculate the reliability of R' and R'' say R''' = R' × R'' = (R1R2R3 + R1R2R4-R1R2R3R4)
Step 4: Calculate the reliability of C5 & R''' in parallel –say R'''' = (C5 + R''' – C5 R''')
Step 5: Calculate the reliability of R'''' and C6 in series

Therefore, the total system reliability R_s = R'''' × R6

= R1R2R3R6 + R1R2R4R6 - R1R2R3R4R6 +R5R6 - R1R2R3R5R6
  - R1R2R4R5R6 + R1R2R3R4R5R6

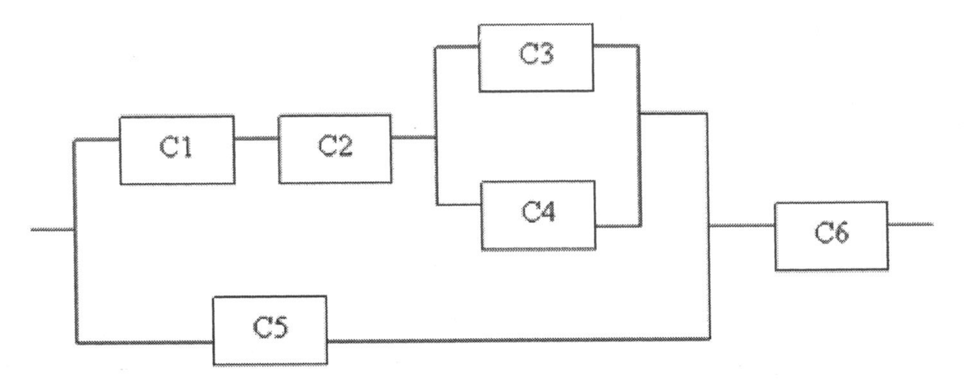

Figure 4-8. Series parallel configuration

EXAMPLE 4.9

Consider the configuration in Figure 4.8. Let the failure time of the components follow exponential distribution. Table 4.3 gives the failure rate of the components and the reliabilities of the components. Find the reliability of the system for a period of 100 hours and also evaluate the MTTF of the system.

Table 4-3. Component reliability information

Component	Failure rate λ / hour	Reliability R(t) (t = 100 hrs)
C1	$\lambda 1 = 0.0004$	0.96
C2	$\lambda 2 = 0.004$	0.67
C3	$\lambda 3 = 0.003$	0.74
C4	$\lambda 4 = 0.003$	0.74
C5	$\lambda 5 = 0.007$	0.50
C6	$\lambda 6 = 0.0006$	0.94

Reliability of the system is given by:

$$R_s(t)=e^{-(\lambda_1+\lambda_2+\lambda_3+\lambda_6)t}+e^{-(\lambda_1+\lambda_2+\lambda_4+\lambda_6)t}-e^{-(\lambda_1+\lambda_2+\lambda_3+\lambda_4+\lambda_6)t}+e^{-(\lambda_5+\lambda_6)t}$$
$$-e^{-(\lambda_1+\lambda_2+\lambda_3+\lambda_5+\lambda_6)t}-e^{-(\lambda_1+\lambda_2+\lambda_4+\lambda_5+\lambda_6)t}+e^{-(\lambda_1+\lambda_2+\lambda_3+\lambda_4+\lambda_5+\lambda_6)t}$$
$$=0.75$$

$$\text{MTTF}_s = \int_0^\infty R_s(t)dt$$

Integrating the expression of $R_s(t)$ above and substituting the values of λ's we get the MTTF_s = 188 hours. In practice the systems resemble configurations as discussed here. We have seen the complexity involved in estimating the MTTF_s. In the example we had just 6 components, where as in practice the components will be replaced with subsystems and each subsystem having several series-parallel combinations of the components. When a system has components in series with time to failure distribution as exponential, both reliability and MTTF_s calculations are fairly simple. But when more parallel combinations are introduced, the reliability and MTTF_s estimation becomes very tedious and complex. Thus one may have to apply certain approximations on the parallel blocks and estimate a failure rate for the parallel blocks so that the MTTF_s estimation can be taken as the reciprocal of the summation of the λ_i's.

4.9 K-OUT-OF-N SYSTEMS

In many practical situations more than one parallel component are required to work satisfactorily for system success. For example, consider a shutdown system consisting of three monitors. To avoid unnecessary shutdown of the system a condition may be imposed that 2 out of 3 generate the signals for the shutdown. Another example can be a car with 6 cylinders which can be driven if at least 4 cylinders are firing, although it might not be a good idea to do so for very long. A k out of n system is represented as shown in Figure 4.9. For deriving the expression for the reliability of such systems: Consider n identical independent components with p as its reliability for a period of time t.

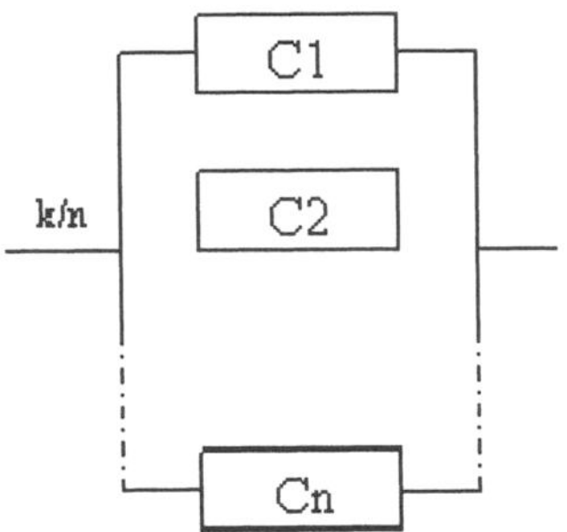

Figure 4-9. K-out-of-N redundancy

The probability, P_r, that exactly *r* failures occur is given by

$$P_r = {}^nC_r \, p^r (1-p)^{n-r} \tag{4.32}$$

For a k out of n system, the system success means that k or, k+1,..., or n components work successfully. So the system reliability will be the sum of the probabilities of k , k+1, ...and n components working successfully. Thus the reliability of k out of n system is given by

$$R_s(t,k,n) = \sum_{r=k}^{n} {}^nC_r \, p^r (1-p)^{n-r} \tag{4.33}$$

For an exponential time to failure distribution the above expression will be:

$$R_s(t,k,n) = \sum_{r=k}^{n} {}^nC_r e^{-r\lambda t} \left(1-e^{-\lambda t}\right)^{n-r} \tag{4.34}$$

4.9.1 Mean time to failure for a K-out-of-N System

The mean time to failure of a K-out-of-N system can be derived from the fundamental equation:

$$MTTF = \int_0^\infty R_s(t)\,dt \tag{4.35}$$

For exponential time to failure distribution

$$MTTF = \int_0^\infty R_s(t)dt = \int_0^\infty \sum_{r=k}^{n} {}^nC_r e^{-r\lambda t}\left(1-e^{-\lambda t}\right)^{n-r} dt \tag{4.36}$$

EXAMPLE 4.10

A power generation system in a factory has six identical generators, each with a failure rate of 1.5 per 1000 hours with the time to failure distribution as exponential. In certain application only four generators need to function. Find the probability of four or more generators functioning for a continuous period of 100 hours. Also find the MTTF of this configuration.

Reliability of one generator for (t = 100hrs) = EXP((-1.5*100)/1000)
= 0.86071

Using equation (4.33) we have

$$R_s(100,4,6) = \sum_{r=4}^{6} {}^6C_r (0.86)^r (0.14)^{(6-r)} = 0.96$$

$$MTTF = \int_0^\infty \sum_{r=4}^{6} {}^6C_r\, e^{-r\lambda t}\left(1-e^{-\lambda t}\right)^{n-r} dt$$

$$= \frac{1}{\lambda}\left(\frac{37}{60}\right) = \frac{1000}{1.5}\left(\frac{37}{60}\right) \approx 411\, hours$$

4.10 STANDBY REDUNDANCY

There are different types of redundancies, which are used for improving the reliability of a system. The one we discussed in section 7 is one type of redundancy where both the components are active and in case of failure of one component the other component takes over the load of the complete function and aids the system functioning without any interruption. This is called active redundancy since both the components are functioning simultaneously. In many applications it is not possible to have an active redundancy. For example the main power supply and uninterrupted power supply (UPS) for a computer cannot be functional simultaneously. The UPS takes over when main power supply fails. This type of redundancy is known as passive redundancy or standby redundancy.

4.10.1 Cold standby system with perfect switching

In a cold standby, the redundant system is switched on only when the main operating item fails. In general the standby configuration is represented as shown in Figure 4.10

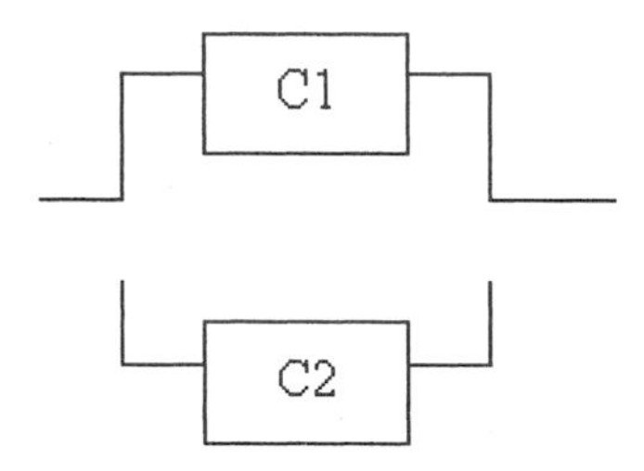

Figure 4-10. Cold standby system

Consider the case where the switching mechanism is 100% reliable. It is also assumed that the standby component does not fail in the standby mode. Under these assumptions the reliability function of the system shown in Figure 4.10 can be derived as follows:

$R_s(t)$ = Prob[The main component C1 survives up to time t] + Prob[C1 fails at time u ($u < t$) and the standby component C2 survives the remaining interval ($t - u$)]

$$R_s(t) = R(t) + \int_0^t f(u)R(t-u)du \qquad (4.37)$$

Where f(t) is the probability density function of the time to failure random variable.

In particular when $f(t) = \lambda e^{-\lambda t}$ we get

$$R_s(t) = e^{-\lambda t} + \int_0^t \lambda e^{-\lambda u} e^{-\lambda(t-u)} du$$
$$= e^{-\lambda t} + \lambda t e^{-\lambda t} = e^{-\lambda t}(1 + \lambda t)$$

For a cold standby system with n identical items with exponential time to failure distribution, the expression for reliability function is given by

$$R_s(t) = e^{-\lambda t} \sum_{i=0}^{n-1} \frac{(\lambda t)^i}{i!} \tag{4.38}$$

It can be observed that the equation (4.38) is the cumulative distribution of Poisson distribution with mean λt. For non-identical items equation (4.37) can be written as

$$R_s = R_1(t) + \int_0^t f_1(x) R_2(t-x) dx \tag{4.39}$$

In case of time to failure following exponential distribution with parameters λ_1 and λ_2 the equation (4.39) will be

$$\begin{aligned} R_s &= e^{-\lambda_1 t} + \int_0^t e^{-\lambda_1 x} e^{-\lambda_2 (t-x)} dx \\ &= e^{-\lambda_1 t} + \frac{\lambda_1}{\lambda_1 - \lambda_2}\left[e^{-\lambda_2 t} - e^{-\lambda_1 t}\right] \end{aligned} \tag{4.40}$$

The MTTF of a cold standby system with identical components can be evaluated by integrating the reliability function in equation (4.38) between 0 and ∞

$$MTTF = \frac{n}{\lambda} \tag{4.41}$$

For non-identical components the MTTF is given by

$$MTTF = \sum_{i=1}^{n} \frac{1}{\lambda_i} \tag{4.42}$$

4.10.2 Cold standby system with imperfect switching

Here we consider that the switching mechanism from the main to standby also has a probability of failure. In that case the reliability function will be as follows, assuming that the probability of the switch functioning is p_s:

$R_s(t)$ = Prob[The main component C1 survives up to time t] +
Prob[C1 fails at time u (u < t) and the standby component C2 survives the remaining interval (t-u) and the successful switching]

$$Rs(t) = R(t) + \int_0^t f(u)R(t-u)p_s\, du \tag{4.43}$$

In case of time to failure following exponential distribution, the above expression will be

$$R_s(t) = e^{-\lambda t} + \int_0^t \lambda e^{-\lambda u} e^{-\lambda(t-u)} p_s du \tag{4.44}$$

The main advantage of a standby redundancy is that it is possible to repair the main unit with uninterrupted operation of the system. The advantage of active redundancy is that the switching mechanism is not required and thus one need not bother about the probability of the switching function.

4.10.3 Hot standby system

In a hot standby the main component and the standby component will be sharing equal load and hence the failure of the standby component can occur even without being operated. The shelf life of the component also plays an important role. The example of UPS can be recalled here, the battery life goes down after certain period. The reliability expression will be the same as parallel redundancy in this case.

4.11 GRACEFUL DEGRADATION

Certain systems have a number of components functioning in parallel, all capable of sharing the load. In such systems failure of a single component will not cause the system failure. For example in signal processing function of a Radar or Sonar there are a number of preamplifiers used to amplify the signals received and do the analysis of a target. In this kind of situation if all the preamplifiers are functional then we get 100% output. In the event of a few amplifiers failing there will be a loss of output. That is we say that the system functions with degraded performance. The degradation or the reduced performance can be represented using some weightage factors. For example let us say there are four preamplifier cards. If all the four function then we have 100% output. Suppose one card fails the system can continue to function with some degradation. The reliability of such a system can be calculated as we did for k out of n systems with a degradation weight added to the terms in the calculation. In such systems fault identifier alarm could be in built. The faulty card could be located and replaced while the system is functioning with degraded performance. After the replacement of the card the system returns to its 100% performance.

EXAMPLE 4.11

There are 4 pre-amplifying cards in the preamplifier of a sonar system. The time to failure distribution of the card is exponential and the failure rate of the card is $15/10^6$ hours. If all the cards function the performance of the system would be 100% satisfactory. If one card fails the performance of the system is reduced by 30%, if two cards fail the performance is reduced by 50%, if three cards fail the performance goes down by 70% and so on. The requirement is that at least 50% performance is expected. Find the reliability of the preamplifier system for a period of 100 hours.

Reliability of one preamplifier card for 100 hours, R(100), is given by:

$$e^{-\frac{15*100}{10^6}} = 0.99850$$

The requirement for accepted 50% performance is that at least 2 cards function.

Reliability of the preamplifier system = P(all the 4 cards functioning) +P(3 cards functioning) + P(2 cards functioning)

$$R_s = {}^4C_4(0.9985)^4 + {}^4C_3(0.9985)^3(0.0015)$$
$$+ {}^4C_2(0.9985)^2(0.0015)^2 = 0.9999$$

The corresponding expected performance is given by:

$$R_s = {}^4C_4(0.9985)^4 + {}^4C_3(0.9985)^3(0.0015)\times 0.7$$
$$+ {}^4C_2(0.9985)^2(0.0015)^2 \times 0.5 = 0.9982$$

That is 99.82%

4.12 CASE STUDY – A SONAR SYSTEM

Sonar is a device for determining through underwater sound, the presence and location of objects in the sea. Various sonar systems exist for underwater navigation and obstacle avoidance. A major classification can be into passive and active sonar. The active sonar system transmits acoustic energy, which is reflected back by the objects under water whereas the passive sonar system listens for acoustic signals emitted by potential targets. Active sonar set is, in its essentials, the underwater equivalent of radar. Its advantage over passive sonar is that it provides a positive method of search, which does not rely on the target behaving in a manner that will expose it to do detection. Sound waves are introduced into the sea by means of a transducer, which can be like a loudspeaker intended to work in water. The sound so created travels along a beam until it encounters a solid body, which causes it to be partially reflected. The partially reflected energy (which is of course an echo), spreads out from the object and some of it returns to impinge on the transducer diaphragm. The pressure on the diaphragm causes electrical voltages at the acoustic frequency, which is then amplified for analysis. The low level signals received are amplified and conditioned by the front-end processor and sent to signal processing modules. The processed signals are sent to display processor for identification and segregation of targets. The processed signals are also sent to audio processors. The identification of targets can be through audio or through visual display. The major subsystems of a sonar system in active mode are described in the block diagram shown in Figure 4.11.

The subsystems shown in the figure are electronics, electromechanical and mechanical in nature. Certain subsystems like Front end processor, Signal processor, display processor are made up of several PCB (printed circuit board) cards and each card having several components. The component details are taken using the failure rates given in MIL-HDBK-217F. The failure rates are taken for naval sheltered environment with the operational temperature 40°C and the failure rate of each card is calculated. The card failure rates are used in calculating the reliability of the subsystems. Table 4.4 lists the subsystems shown in the RBD with their failure rates and reliability calculated for 100 hours of continuous operation. The time to failure distribution of the subsystems is exponential. The parallel configurations shown in the RBD have their reliabilities calculated as per equation (4.20). After evaluating the parallel configurations the subsystem's reliabilities can be multiplied as in series system (equation 4.5) and we get the reliability of the sonar system to be 0.8013.

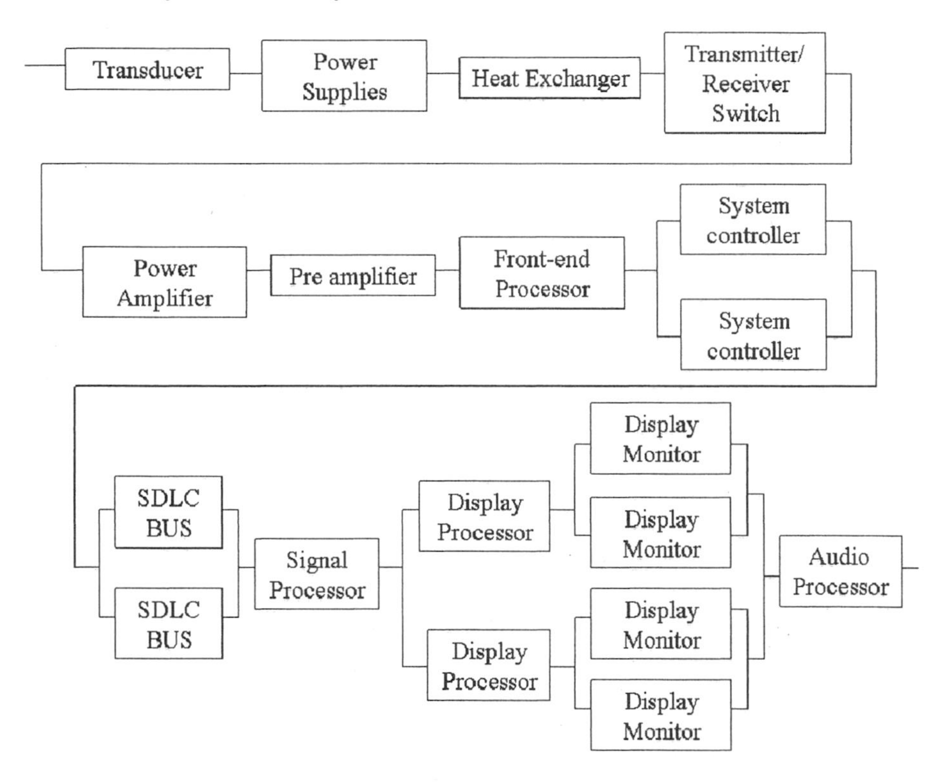

Figure 4-11. Reliability block diagram of a SONAR system

The audio processor in the block diagram is not a critical subsystem as far as the functionality of the sonar is concerned. Audio and Display are both for identifying the targets. Certainly we cannot consider audio and display in parallel since display system gives more information on the targets identified. However, the identification of targets by audio is preferred in certain cases. Thus the above calculated reliability of the system is with the audio present. The reliability without the audio will be 0.8025.

$$\text{The MTTF of the sonar system} = \int_0^\infty R_1(t) \times R_2(t) . \times \cdots \times R_{15}(t)\, dt$$

Substituting the expression for the reliability of each subsystem in the above integral the MTTF = 472 hours. The reliability calculation is as per the block diagram using the series and parallel configurations shown. For 100% satisfactory performance, all the subsystems and the PCB cards present in the subsystem should be functional. There are certain subsystems which will still be working even when some component/PCB cards fail, but with a degraded performance without stalling the total performance of the system. The preamplifier and power amplifier have been provided with such graceful degradation.

The power amplifier and preamplifier have four cards. If all the four cards function then we have 100% performance. If three cards function the performance is degraded by 20% and if only two cards function performance is degraded by 50%. Thus if we accept the system performance up to 50% degradation then the system reliability can be recalculated with the respective failure rates of the cards present in the system similar to the example (4.11).

Under the accepted degraded performance, the reliability of

Power amplifier = 0.99755

Preamplifier = 0.999358

The reliability of the sonar system with these acceptable level of graceful degradation is equal to 0.811052 (with the audio processor present). Several software packages are available for drawing the RBDs and to carry out the reliability calculations. Some of them are: ITEM® software, RELEX® software and ReliaSoft®. Every software has its own limitations, especially with respect to the database of the failure rates of the components. With the advancement in technology we have more and more custom made components, which we may not find in the database of these software. Hence it is essential that the failure rates for such components are determined using field data and other accelerated test methods.

Table 4-4. Failure rates of subsystems in a SONAR

Sl. No.	Subsystem	Failure rate	Failure rate $\lambda/10^6$ hours	Reliability For T=100hrs
1	Transducer	$\lambda 1$	1000	0.904837
2	Power Supplies	$\lambda 2$	20	0.998002
3	Heat Exchanger	$\lambda 3$	20	0.998002
4	Transmitter/Receiver Switch	$\lambda 4$	6	0.999400
5	Power Amplifier	$\lambda 5$	120	0.988072
6	System Controller (parallel)	$\lambda 6$	10	0.995999
7	SDLC Bus (parallel)	$\lambda 7$	15	0.993998
8	External System Interface	$\lambda 8$	30	0.997004
9	External Classifier	$\lambda 9$	10	0.999
10	Preamplifier	$\lambda 10$	32	0.996805
11	Front End Processor	$\lambda 11$	400	0.960789
12	Signal Processor	$\lambda 12$	450	0.955997
13	Display Monitor (parallel)	$\lambda 13$	50	0.979975
14	Display Processor (parallel)	$\lambda 14$	150	0.998802
15	Audio processor/Audio	$\lambda 15$	20	0.998501

Chapter 5

DESIGN FOR RELIABILITY AND SIX SIGMA

Whenever anyone says, 'theoretically', they really mean, 'not really.'
Dave Parnas

5.1 INTRODUCTION

Six-sigma is basically a goal of process quality. However, it must always be remembered that a quality process is just one of the key elements of product reliability. Designing for reliability encompasses a great deal more than this. If the individual parts themselves do not start with an adequate level of quality, and if the need for reliable design is not adequately addressed during the development, final product reliability at the six-sigma level is simply not possible. Design for six-sigma is a business process focused on improving profitability when properly applied. It generates the right product or Service at the right time at the right cost.

One of the important tasks of the design process is to translate the overall functional requirements for a new system into its physical requirements in relation to the performance, power consumption and cost. In this book we have discussed six sigma and reliability. Reliability ensures failure free design and Six Sigma ensures defect free process. Implementation of six-sigma program requires close co-operation between the product design effort and the design of the manufacturing process. One of the goals of product design is to minimize the variations in the process to ensure the production of the desired characteristics and to center the process on the target nominal value of the desired parameter.

There are several analysis tools for building reliability and ensuring a six sigma process in the design. The main design analysis tools are

- Quality function deployment (QFD)

- Reliability allocation
- Accelerated testing
- Highly accelerated life testing
- Safety Analysis
- Failure reporting and corrective action system (FRACAS)

5.2 QUALITY FUNCTION DEPLOYMENT

Quality function deployment is also known as House of Quality. The name house of quality comes from the fact that the analysis chart matrix is in the shape of a house. Yoji Akao is widely regarded as the father of QFD and his work led to its first implementation at the Mitsubishi Heavy Industry's Kobe Shipyard in 1972. Yoji Akao considered QFD as a method for developing a quality design aimed at satisfying the customer and then translating the consumer's demands into design targets and also to implement the major quality assurance points to be used throughout the production phase. The main focus of QFD is meeting customer needs through the use of their actual statements (known as voice of customer, VOC). The QFD translates the customer requirements, the market research and technical benchmarking data into an appropriate number of prioritised engineering targets to be met by a new product design.

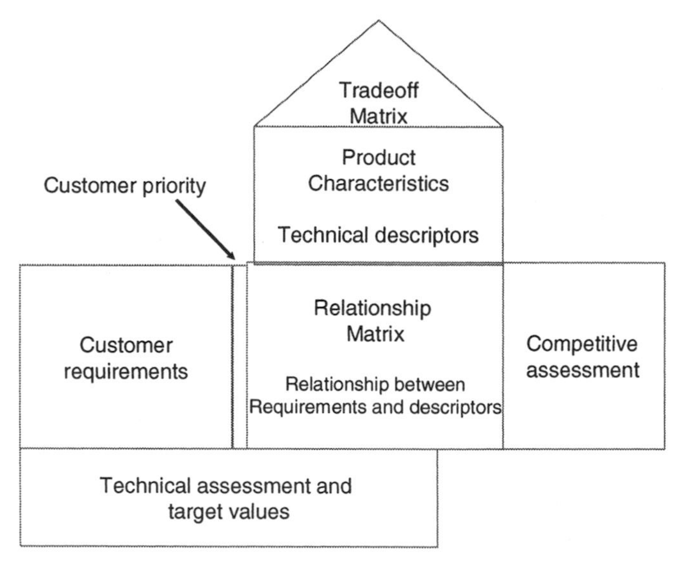

Figure 5-1. QFD Matrix

Quality Function Deployment (QFD) translates decision criteria or Critical-To-Quality (CTQ) parameters into a prioritized set of targets, choices, or improvement opportunities - helping to produce better products, processes, services, or strategies.

The exterior walls of the house are the customer requirements (Figure. 5.1). On the left side is a list of the voice of the customer, which are the customer requirements. Immediately after customer requirements are the prioritized customer requirements, or planning matrix (this can be placed just before customer competitive assessment as shown in Figure 5.2). On the right side of the QFD matrix is the customer competitive assessment of customer benchmarking. The ceiling or the second floor of the house contains the technical descriptors or product characteristics. Consistency of the product is provided through engineering characteristics, design constraints and parameters. The interior walls of the house are the relationship between customer requirements and the technical descriptors. Customer requirements are translated into engineering characteristics (technical descriptors). The roof of the house is the correlation matrix which describes the interrelationship between technical descriptors. Tradeoffs between similar and/or conflicting technical descriptors are identified. The foundation of the house is the prioritized technical descriptors. Items such as technical benchmarking, degree of technical difficulty and target value are listed.

5.2.1 Building house of quality matrix

In this section, we describe the steps involved in building a QFD matrix.

Step 1 - List customer requirements (CR)

The customer requirements are captured by conversations with the customer in which they are encouraged to state their needs, preferences and problems. The captured data is arranged in a structural form such as performance requirements, usability requirements and aesthetic requirements. To illustrate the steps let us consider an example shown in Figure 5.2. The example describes a product, Rock climbing Harness. The customer requirements are listed in the figure to the left of the house.

Step 2 – List the technical descriptors (TD)

The QFD team comes up with the engineering characteristics or technical descriptors that will affect one or more of the customer requirements. These technical descriptors make up the ceiling of the house. In the example we can see that harness weight, webbing strength and padding strength are some of the technical descriptors. The technical descriptors can be classified and grouped; such as performance measure, size of range and technical details.

Step 3 – Develop a relationship matrix between CR's and TD's

Now we compare the customer requirements and the technical descriptors and determine their respective relationships. We fill these columns with certain symbols. To reduce the complexity we use only three different grading such as strong, medium and weak interrelationship. The symbols are shown with their identification of importance in the figure 5.2. These grading can also be assigned some value say 9 for strong relationship, 3 for medium and 1 for weak. The relationship matrix infers the degree of technical descriptors on the customer requirements.

After the interrelationship matrix has been completed, it is evaluated for empty rows or columns. An empty row indicates that a customer requirement is not being addressed by any of the technical descriptors. This will indicate that additional technical descriptors have to be designed for satisfying the customer requirement. An empty column indicates that a particular technical descriptor is not important in meeting any of the customer requirements and after careful study one can remove it from the house of quality. The example shows the interrelationship filled in with various symbols.

Step 4 – Develop an interrelationship Matrix between TD's

The roof of the house, also known as the correlation matrix is the interrelationship between the technical descriptors. The positive sign (+) indicates a positive correlation and the negative sign (-) indicates conflicting technical descriptors. This helps in identifying which technical descriptors support one another and which are the ones in conflict. The conflicting technical descriptors are extremely important since they are frequently the result of conflicting customer requirements. This helps in figuring out where the trade offs should be targeted in satisfying the customer. Tradeoffs that are not identified and resolved will often lead to unfulfilled requirements, engineering changes, increased costs and poorer quality. Some of the tradeoffs may require high level managerial decisions because they cross functional area boundaries. An example of a tradeoff in the design of a car is the conflict between the customer requirements of high fuel economy and safety related technical descriptors. The added weight of stronger bumpers, airbags, antilock brakes ultimately reduce the fuel efficiency of the car.

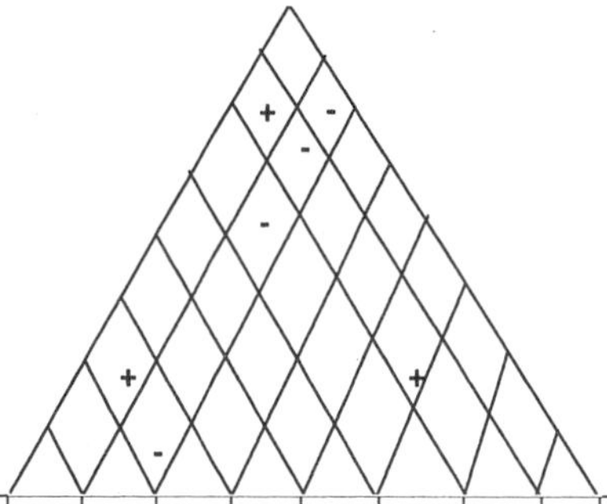

Technical Descriptors→ / Customer Requirements ↓	Meets Euro Standards	Harness Weight	Webbing Strength	No. of colors	No. of sizes	Padding thickness	No. of buckles	No.of gear loops	Customer importance	Our product	A's product	B's Product	Target value	Scale up factor	Sales point	Absolute value
Easy to put on					■		●		2	3	3	4	4	2	1.1	4.4
Comfortable when hanging					■	●	■		5	4	4	2	5	1	1.4	7
Fits over different clothes					■	■	●		1	1	1	5	2	2	1.0	1
Accessible gear loops								●	3	3	4	1	3	1	1.0	1
Does not restrict movement		■			■	●	■		5	2	2	3	5	1	1.4	7
Light weight		●	■			■	▲	▲	3	3	2	5	3	1	1.0	3
Safe	●	■	●						5	4	3	3	4	0.8	1.2	4.8
Attractive		▲		●		▲	▲		2	2	2	5	3	1.5	1.1	3.3
Our product	5	5	5	5	4	5	1	4								
A's product	5	3	3	3	5	2	4	5								
B's product	5	5	4	6	4	4	1	3								
Degree of difficulty	1	3	1	3	2	1	3	1								
Target value	5	5	1	4	4	1	4	1								
Absolute weight	45	59	54	18	39	104	62	30								
Relative weight	43.2	65.7	29.7	52.2	58.2	141.3	96.9	30								

▲ Strong interrelationship (9)

■ Medium interrelationship (3)

● Weak interrelationship (1)

Figure 5-2. QFD matrix for Rock Climbing Harness

(Source: http://www.shef.ac.uk/~ibberson/Images/QFD/ Reproduced from website with permission from Professor K Ridgway)

Step 5- Competitive assessments

The competitive assessments are of two kinds: Customer assessment and technical assessment as shown in the Figure 5.1. Numbers 1 through 5 are used in competitive evaluation column to indicate a rating of 1 for the worst and 5 for the best.

Customer competitive assessment: This qualifies the customer requirements priorities and their perceptions of the performance of existing products. Secondly it allows these priorities to be adjusted based on the issues that concern the design team. To do this first we need to have the *importance weighting* for the customer requirements. This quantifies the relative importance of each of the customer requirements described from the customer's own perspective. Next we rate our product with respect to various customer requirements and also compare it with that of other manufacturers. A questionnaire can be used to gather these importance-weightage (Say 5 highly important - 1 least important). This can be filled from market sales and survey of other products. This also gives an assessment of where the current product stands in relation to the other competitors in the market.

Technical competitive assessment: This is filled in the same way with ratings 1 (low) to 5(high) for each of the technical descriptor. These make up the rows at the bottom of the house beneath the relationship matrix. The technical competitive assessment is useful in filling up the gaps in engineering judgment. When a technical descriptor directly relates to a customer requirement, a comparison is made between the customer's competitive evaluation and the objective measure ranking. Customer requirements and technical descriptors that are strongly related should also exhibit a strong relationship in their competitive assessment and vice a versa.

Step 6 – Develop prioritized customer requirements

The prioritized customer requirements make up a block of columns corresponding to each customer requirement in the house of quality on the right side of the customer competitive assessment as shown in the example. This has the columns importance to the customer, target value; scale up factor, sales point and absolute weight.

Importance to customer

We can use grading numbers say 1 through 5 to indicate the rating of 1 for least important and 5 for very important. Importance ratings represent the relative importance of each customer requirement in terms of each other. Assigning ratings to customer requirements is sometimes difficult, because each member of the QFD team might believe different requirements should be ranked higher. The importance rating is useful for prioritizing efforts and making trade-off decisions. For example, 'does not restrict movement' has a weight 5 as it is considered more important in comparison to 'fits over different clothes' which has a weight 1.

Target value

The target value column is on the same scale as the customer competitive assessment (1 for worst, 5 for best can be used). This column is where the QFD team decides whether they want to keep their product unchanged, improve the product, or make the product better than the competition.

Scale-up-factor

The scale up factor is the ratio of target value to the rating of the product given in the customer competitive assessment. The higher the number, the more effort is needed. This tells us where the current product is and whether we are reasonably close to target value. If the difference is too much then, may be, we have to have a re-look at target value. In the example we can see that the scale up factor varies between 0.8 and 2. The requirement 'easy to put on' has a higher value over 'fits over different clothes' indicating a greater effort needed to achieve it.

Sales point

The sales point tells the QFD team how well a customer requirement will sell. The objective here is to promote the best customer requirement and any remaining customer requirements that will help in the sale of the product. For example, the sales point is a value between 1 and 2, with 2 being the highest. In the example 'comfortable when hanging' may give a better sales opportunity and thus takes a value 1.4. If a customer requirement does not help the sales then it is given a value 1. In the example 'fits over different clothing' is given a value 1.

Absolute weight

Absolute weight is calculated by taking the product of the importance to customer, scale-up factor and sales point.

$$\text{Absolute weight} = (\text{importance to customer}) \times (\text{scale up factor}) \times (\text{sales point}) \quad (5.1)$$

A sample calculation from the example can be absolute weight for *comfortable when hanging* = (5 x 1 x 1.4) = 7.0

Step 7 – Develop prioritized technical descriptors

These are the rows corresponding to the technical descriptors described on the top of the tree. The prioritized technical descriptors contain the technical priorities, target value, absolute weights and relative weights. The most important technical descriptors required for fulfilling the customer requirements are identified. This measure provides specific objectives that guides the subsequent design and provide a means of objectively assessing progress and minimizing subjective opinions.

The first three rows describe the rating of technical specifications met by our product and also the other products in the market. Then we have the Degree of difficulty stating how easy it is to meet the technical specifications. (1 indicates least difficulty and 5 indicate most difficult). Then we calculate the absolute weight and relative weight with respect to the technical descriptors. The absolute weight for the technical descriptors can be calculated as follows:

$$T_j = \sum_{i=1}^{n} R_{ij} C_i \quad (5.2)$$

Where T_j is the jth technical descriptor, R_{ij} is the relationship matrix elements and C_i is the customer importance column. For example the technical priority value for harness weight in Figure 5.2 will be:

T_2 = (3 x 5) + (9 x 3) + (3 x 5) + (1 x 2) = 59

In the given example we see that padding thickness has a greater impact on the design of the rock climbing harness.

Relative weight

The relative weight of the jth technical descriptor is calculated by replacing the customer importance column by the absolute weight of the customer requirements. It is calculated as follows:

$$a_j = \sum_{i=1}^{n} R_{ij} d_i \tag{5.3}$$

Where:
a_j = row vector of the relative weights for the technical descriptors (j = 1 ... m)
d_i = column vector of the absolute weight of the customer requirements (i= 1 ... n)

Thus, the relative weight for the padding thickness is

$a_2 = (3 \times 7) + (9 \times 3) + (3 \times 4.8) + (1 \times 3.3) = 65.7$

Higher absolute and relative weights indicate areas where engineering efforts need to be concentrated. The primary difference between these two weights is that the relative weight also includes information on customer scale up factor and sales point. Also these weights show the impact of the technical characteristics on the customer requirements. Using relative weights the designer can identify the technical descriptor necessary for meeting the customer requirements. Along with the degree of technical difficulty decisions can be made concerning where to allocate resources for quality improvement.

Like every tool QFD also has its limitations. To get the maximum benefit out of the analysis, it should be supplemented with other related reliability analyses tools such as FMECA, FTA and safety analysis for the technical descriptors before implementation. Before we add a technical descriptor or delete one from those listed based on the analysis, care should be taken to preserve the basic functionality of the product. In general many technical descriptors are added to meet the customer requirements. But it is very important to refine the technical descriptors till an actionable level of detail is achieved. Thus at times the QFD may have to be repeated in an iterative process. The process should be repeated till each objective is refined to an actionable level.

EXAMPLE 5.1

House of quality matrix has been constructed for the product – Rock climbing Harness and the matrix is shown in Figure 5.2.

5.3 RELIABILITY ALLOCATION

In a large complex system, it is necessary to translate reliability requirements into detailed specifications for the numerous units that make up the system. The overall system reliability requirement is translated or broken down into subsystem, component or part reliabilities. This is essential in particular when different design teams, subcontractors or sub-manufacturers are involved. The appropriate values of reliability have to be included in the subsystem, component, or part specifications.

The process of apportioning reliability requirement to individual subsystems, components or parts from specified system reliability is called Reliability Allocation or Reliability Apportionment. This allocation technique varies with the complexity of the system and the role of the subsystem, components, or parts in the functioning of the system. The sensitivity of the reliability allocation at the design stage to the uncertainty of its parameters is a crucial consideration. Usually, weight, the initial cost of the system or performance characteristics are used as objective functions. If the allocated parameters for a system cannot be achieved using current technology or is not cost effective then the allocation should be reassigned. A trade-off analysis among the contradicting parameters should be made and the procedure should be repeated until an allocation is achieved that satisfies the reliability requirements.

5.3.1 Mathematical formulation

The allocation of system reliability involves solving the basic inequality:

$$f(R_1^*, R_2^*, ..., R_n^*) \geq R^* \tag{5.4}$$

Where

R^* is the required system reliability

R_i^* is the ith subsystem reliability requirement

For a series system, equation (5.4) is simplified as:

$$R_1^*, R_2^*, \ldots, R_n^* \geq R^* \tag{5.5}$$

Theoretically this equation has an infinite number of solutions, assuming that there is no restriction on the allocation. The problem is to establish a procedure that yields a solution by which consistent and reasonable reliabilities may be allocated. There are various approaches to reliability allocation and some of them are listed below:

1. Equal apportionment technique
2. AGREE apportionment technique
3. ARINC apportionment technique
4. Feasibility of objectives technique
5. Dynamic programming approach
6. Minimum effort algorithm

Of the several techniques mentioned above the equal apportionment technique and minimum effort algorithm are described in this chapter. Equal apportionment technique is used when we do not have any information about the subsystems' reliability. Minimum effort algorithm is used when we have a combination of new subsystems and a few subsystems with known reliability.

5.3.2 Equal apportionment technique

In the absence of definite system information, other than the fact that it contains **n** subsystems in series, equal apportionment to each subsystem would seem reasonable. In this case, the nth root of the system reliability requirement would be apportioned to each subsystem. Equal apportionment assumes a series of n subsystems, each of which is to be assigned the same reliability goal. The technique is based upon the formula:

$$R_i = (R_{system})^{\frac{1}{n}} \tag{5.6}$$

where:

R_i = reliability allocated to the ith item

R_{system} = required system reliability

n = total number of items

The primary weakness of this method is that the subsystem goals are not assigned in accordance with the degree of difficulty associated with achievement of these goals. This does not take into consideration complexity and other design factors, and hence extensive reallocation would probably be necessary.

EXAMPLE 5.2

A system has 3 subsystems connected in series with the overall reliability requirement of 0.95. Allocate the reliability to the subsystems using equal apportionment technique.

Given $R_{system} = 0.95$

$$R_1 = R_2 = R_3 = (0.95)^{\frac{1}{3}} = 0.983$$

5.3.3 Minimum effort algorithm

Let us consider a system with m subsystems and the reliability goal for the system is set as R*. The subsystems have reliabilities say R_1, R_2 , ..., R_m. In order to achieve the reliability goal, R*, we need to improve the reliability of at least a few of the subsystems where ever possible. To improve the reliability of any system, efforts are needed in terms of using better components and/or conducting more tests to eliminate major failures of the subsystems. This method addresses a way in which minimum effort is required to increase the reliability.

Let the effort function $G(R_i, R_i^*)$ be a measure of the amount of effort needed to increase the reliability of the ith subsystem from Ri to Ri* . The assumption is that the nature of effort function is same for all subsystems. In this method we find out for how many subsystems we can improve the reliability and the value to which it can be increased. The reliabilities of rest of the subsystems are left unchanged. The following steps are used in achieving this task.

1. The estimated or predicted reliabilities are arranged in ascending order. Without loss of generality, assume that:

$R_1 \leq R_2 \leq ...... \leq R_m$

2. The Reliabilities of first **k** components are increased to the value R_0. The reliabilities $R_{K+1}, R_{K+2}, \ldots R_m$ of the remaining (m-k) subsystems are left unchanged.

3. The number **k** is the maximum value of j such that:

$$R_j < \left[\frac{R^*}{\prod_{i=j+1}^{m+1} R_i} \right]^{1/j} = r_j \tag{5.7}$$

Where $R_{m+1} = 1$. For illustration, first we set the value of r_1 using equation (5.7) as shown below:

$$\left[\frac{R^*}{R_2 R_3 \ldots R_m} \right]^{1/1} = r_1$$

If $R_1 < r_1$ then R_1 will have to be increased to the value of R_0 (equation 5.8). Then, we calculate the value of r_2 which is given by:

$$\left[\frac{R^*}{R_3 R_4 \ldots R_m} \right]^{1/2} = r_2$$

If $R_2 < r_2$, then both R_1 and R_2 will have to be increased to a common value, R_0. The procedure is continued as long as $R_j < r_j$. The value R_0, the allocated reliability for first k units is determined using equation (5.8).

$$R_0 = \left[\frac{R^*}{\prod_{i=k+1}^{m+1} R_i} \right]^{1/k} \tag{5.8}$$

EXAMPLE 5.3

A system consists of four subsystems connected in series. $R_1 = 0.7$, $R_2 = 0.85$, $R_3 = 0.92$, and R_4=0.9. The overall system reliability goal is set to 0.62. It is given that the improvement of reliability of subsystem 4 is not possible. Using the minimum effort method, allocate the reliabilities to the subsystems.

$R^* = 0.62$

It is given that reliability improvement is not possible on subsystem 4 and hence R4 is fixed and cannot be changed. Therefore the reliability goal for the rest of the three subsystems will be

$$R' = (R^* / R_4) = (0.62 / 0.9) = 0.69$$

Here $R_1 < R_2 < R_3$

$$r_1 = (0.69 / (0.85*0.92) = 0.882 > 0.7$$
$$r2 = ((0.69 / (0.92))^{1/2} = 0.866 > 0.85$$
$$r3 = (0.69)^{1/3} = 0.884 < 0.92$$

The minimum effort method suggests that the reliability of the subsystem 1 & 2 can be improved. The value R_0 to which the subsystems 1 and 2 can be improved is

$$R_0 = (0.69 / 0.92)^{1/2} = 0.866$$

Thus $R_1^* = R_2^* = 0.866$, $R_3 = 0.92$ and $R_4 = 0.9$

$$(R_1^*)(R_2^*)(R_3)(R_4) = 0.621$$

5.3.4 Apportionment for new units

Very often reliability is to be allocated to subsystems for which no estimated/predicted reliability values are known. It can be logically done keeping in view factors like complexity, cost factor, redundancy factor, maintainability factor, state of technological art factor, duty ratio. In this section, a series logic diagram is assumed. If the modules are connected in parallel in any subsystem; the parallel unit is considered as one entity for the purpose of reliability allocation. This approach allows us to retain the validity of series structure.

Let there be **N** subsystems in the system whose reliability goal is R^*. Out of these N subsystems, let there be $m(\leq N)$ subsystems whose estimated/predicted reliabilities are known and reliability improvements are considered feasible. Let n = N - m be the remaining subsystems whose estimated/predicted reliabilities are not known and we have to allocate reliabilities to these subsystems considering parameters such as cost, complexity, state of art, redundancy, maintainability, duty ratio and criticality of the system.

This problem could be decomposed into two independent reliability allocation problems involving m and n units respectively. For the first m components the goal is taken as $(R^*)^{\frac{m}{N}}$ and for the remaining n components in the second category, the goal is taken as $(R^*)^{\frac{n}{N}}$. For simplicity of notation let:

$$\bar{R} = \left(R^*\right)^{\frac{m}{N}} \tag{5.9}$$

and

$$\hat{R} = \left(R^*\right)^{\frac{n}{N}} \tag{5.10}$$

The reliability allocation to the m units could be carried out as described by equations 5.7 and 5.8 (that is using minimum effort algorithm). The reliability allocation to the new n units can be carried out as described below.

Apportion the reliability goal $\hat{R}$ to **n** units such that:

$$\prod_{j=1}^{n} R_j^* > \hat{R} \tag{5.11}$$

If λ_j^* is allocated failure rate for j[th] subsystem and $\hat{\lambda}$ is the required failure rate for the system, equation (5.11) is equivalent to:

$$\sum_{j=1}^{n} \lambda_j^* \leq \hat{\lambda} \tag{5.12}$$

As λ_j^* is obviously a fraction of the total failure rate, let

$$\lambda_j^* = W_j \hat{\lambda} \tag{5.13}$$

where W_j is the weightage factor for j^{th} subsystem. These weightage factors have to be defined such that:

$$\sum_{j=1}^{n} W_j = 1 \tag{5.14}$$

Also, R_j^* can be expressed as:

$$R_j^* = \left(\hat{R}\right)^{w_1} \tag{5.15}$$

To make sure that equation (5.14) is satisfied, we define W_j in terms of proportionality factors Z_j's as:

$$W_j = \frac{Z_j}{\sum_{j=1}^{n} Z_j} \tag{5.16}$$

The proportionality factor Z_j for j^{th} subsystem is defined in terms of various factors based on which desired reliability is allocated. A discussion of these factors follows in the next section. It may be noted that higher Z_j implies higher W_j which implies higher allocated failure rate and hence lower allocated reliability.

5.3.5 Reliability allocation factors

Reliability allocation depends on several factors and some of these factors are discussed in the following sections.

5.3.5.1 Complexity factor

Complexity of a system is the number of components comprising it. Higher the complexity of the system, higher is the complexity factor. Reliability of any module depends upon the number of components comprising it; therefore, reliability allocation should have a strong dependence on complexity. It is known that the failure rate of any module is the sum of the failure rates of the constituent components (for a series system). As a first approximation, therefore, it is logical to have allocated failure rate of any module proportional to the number of components comprising it. Hence,

$$Z_j \propto K_j \tag{5.17}$$

Where, K_j is the complexity factor for j^{th} subsystem. These complexity factors are generally measured as the number of Active Element Groups, AEG's.

5.3.5.2 Cost

The improvement in the reliability of some subsystems may involve larger increments in cost. Therefore, lower value of reliability improvement should be allocated to an expensive subsystem. The apportioned reliabilities are supposed to be demonstrated and demonstration of a high reliability value for a costly system may be extremely uneconomical. From this point of view also, a relatively lower value of reliability is desired to be allocated to a costlier subsystem. As higher failure rate is desired to be allocated to a costlier system,

$$Z_j \propto C_j \tag{5.18}$$

Where, C_j is the cost for j^{th} subsystem.

5.3.5.3 State of the Art

If a component has been available for a long time and has experienced an extensive development program including failure analysis and corrective action for deficiencies, it may be quite difficult to further improve its reliability even if the reliability is considerably lower than desired. Other components which have initially higher reliabilities may be further improved more economically.

As state of art is the possibility of achieving improvement, the value of this factor is an engineering judgement. When no improvement can be achieved, a factor of 1 is chosen and for the subsystem where lot of improvement is possible, a larger value, S_j is chosen. Obviously larger S_j means higher reliability and hence lower failure rate. Therefore,

$$Z_j \propto \frac{1}{S_j} \tag{5.19}$$

5.3.5.4 Redundancy factor

Redundancy increases the reliability considerably, often dramatically. It can be shown for example that if two systems, each with reliability of 0.8 are connected in parallel to provide active redundancy, then the combination will have a reliability of 0.96. The increase in the reliability value depends on the reliability values of the individual subsystems and the number of units connected in parallel. There are two ways in which this increase in reliability of a redundant system can be made use in practice. One, using redundancy, it is possible to achieve values of reliability which are difficult or impossible to get in practice with a single system. Second, it is possible to use subcomponent systems of considerably lower reliability and lesser cost in active redundancy. The use of standard components in such systems will help to keep the cost down while increasing reliability of the combination. The factor α can be chosen accordingly. Higher the value of α implies allocating reliability corresponding to higher failure rate. If the redundancy is not possible α is set equal to 1

$$Z_j \propto \alpha_j \tag{5.20}$$

5.3.5.5 Maintainability factor

Any system that is periodically repaired and maintained will have a higher availability. Hence a lower reliability can be apportioned to subsystems where maintenance is possible. The quantification of this factor is also an engineering judgement. For non-maintainable subsystems a factor of 1 is chosen and for the subsystems where maintenance is possible, a larger value, M_j is chosen. As stipulated, larger M_j implies higher availability enabling us to apportion reliability corresponding to a higher failure rate. Therefore,

$$Z_j \propto M_j \tag{5.21}$$

5.3.5.6 Duty ratio factor

The duty ratio factor for a subsystem can be defined as the function of the mission time for which it operates. If T is the mission time and also the operating time of all subsystems, time of operation need not be considered in reliability allocation. However, for a sophisticated mission it is quite probable that some subsystems are required to operate for periods less than the mission time. Relatively lower value of reliability is apportioned to the subsystem whose operation time is less than the mission time. Hence,

$$Z_j \propto \frac{1}{d_j} \tag{5.22}$$

$$d_j = \frac{t_j}{T} \tag{5.23}$$

Where d_j can be defined as the duty ratio for j^{th} subsystem, that is, the fraction of the mission time for which j^{th} subsystem operates.

5.3.6 Reliability allocation formula

After considering all the above mentioned factors, the formula for reliability allocation can be expressed as follows:

Combining equations (5.17) through (5.23)

$$Z_j = \frac{K_j C_j \alpha_j M_j}{S_j d_j} \tag{5.24}$$

Proportionality sign has been replaced by equality without any loss of generality, as any constant will cancel out when equation (5.16) is used for computation of weightage factors. To choose the factors in equation (5.24) a scale of 1 to 10 may be appropriate for most of the situations. If any factor in equation (5.24) is considered insignificant for a particular system or if any factor is not valid in a particular case, this can be easily incorporated in the formula by taking unit value for such factors for all the subsystems. Similarly, if desired, a new factor can be incorporated in equation (5.24). For example criticality factor for a subsystem could be introduced. Criticality of a subsystem is the probability of the system failure if the particular subsystem fails. Similarly financial risk factor could be introduced. It depicts the financial loss involved when the system's first stage fails or the last stage fails. Subsystems which have high financial risk should be allocated high reliability value.

After calculating Z_j's for all stages, the weightage factors can be obtained using (5.16) and the values of apportioned reliabilities using (5.15).

EXAMPLE 5.4

Consider a system having 10 subsystems whose required reliability is 0.70. The estimated reliabilities of first five subsystems are 0.95, 0.90, 0.94, 0.96 and 0.98 respectively. The reliabilities of the first two of the first five subsystems cannot be improved, while remaining three are available for possible improvement. The estimated reliabilities of the last five subsystems are not known but the following data are known about these:

1. Subsystems 7 and 8 operate for 75% and 50% of the mission time respectively. All other subsystems operate for complete mission time.
2. Redundancy can be used at subsystems 6 and 10 only.
3. Maintenance is not possible for any of the subsystems.
4. The values of complexity factor, cost factor and state of art factor are known for all the five systems and are listed in the following table.

j	K_j	C_j	α_j	S_j
6	2	2	2	1.0
7	5	2	1	4.0
8	3	2	1	3.0
9	7	4	1	5.0
10	2	2	2	2.0

Reliability is to be allocated for the subsystems of this system.

Subsystems 1 and 2 have their estimated reliabilities known which cannot be improved further. Therefore, we can take these two subsystems out of the purview of reliability allocation by modifying the reliability goal as:

$$R^* = \frac{(0.70)}{(0.95)(0.90)} = 0.819$$

By using (5.9) and (5.10) the reliability goals for subsystems 3 to 5 and for subsystems 6 to 10 are established as:

$$\hat{R} = \left(R^*\right)^{\frac{m}{N}} = (0.819)^{\frac{3}{8}} = 0.928$$

$$\overline{R} = \left(R^*\right)^{\frac{n}{N}} = (0.819)^{\frac{5}{8}} = 0.883$$

First consider the problem of reliability allocation to the first group of systems. Using (5.7):

$$r_3 = \left[\frac{0.928}{(0.96)(0.98)}\right] = 0.986 > 0.94$$

$$r_4 = \left[\frac{0.928}{0.98}\right]^{\frac{1}{2}} = 0.973 > 0.96$$

$$r_5 = \left[\frac{0.928}{1}\right]^{\frac{1}{3}} = 0.978 < 0.98$$

$$R_0 = \left[\frac{0.928}{0.98}\right]^{\frac{1}{2}} = 0.973$$

Therefore subsystem 5 need not be improved while subsystems 3 and 4 can be improved to R_0 using relation (5.8).

Hence

$$R_3^* = R_4^* = 0.973 \quad and \quad R_5^* = 0.980 \quad (unchanged)$$

For reliability allocation to subsystems 6 -10, we use the method outlined in section (3.5). We first calculate the proportionality factors using (5.24).

The Z_j calculation and its factors are shown in the following table:

j	K_j	C_j	S_j	α_j	M_j	d_j	Z_j calculated	W_j
6	2	2	1	2	1	1	8	.3007
7	5	3	4	1	1	0.75	5	.1880
8	3	2	3	1	1	0.5	4	.1504
9	7	4	5	1	1	1	5.6	.2105
10	2	2	2	2	1	1	4	.1504
							Sum=26.6	

$$Z_j = \frac{K_j\ C_j\ \alpha_j M_j}{S_j\ d_j}; \qquad W_j = \frac{Z_j}{\sum Z_j}$$

Hence the allocated reliabilities to these subsystems are given by

$$R_j^* = \hat{R}^{W_j} \quad \text{for } j = 6 \text{ to } 10 \text{ (using equation (5.15))}$$

5.4 ACCELERATED TESTING

In order to assure reliability of a product the design of the product should be reliable which performs its functionality under the required environmental conditions. One of the conventional methods of assuring this has been by analyzing the time to failure data obtained under normal operating conditions in order to quantify the reliability or the MTBF demanded by the customer. With the advance in technology and the competition in the market it is really difficult to capture the time to failure data under the normal operating conditions. Also one cannot capture all the hidden failure modes which may occur during the life span of the product. To assure the reliability one needs to analyze all the failures in the least possible time so that the product could be launched in the market without losing to competition. The accelerated test aims at testing the product for all its latent failures and implementing the required design changes to assure a fault free operation of the product to the extent possible in the real world.

A variety of methods which serve different purposes have been termed as accelerated life testing. They involve quantification of the life characteristics of the product at normal use conditions. The accelerated tests can be divided into two categories, namely qualitative tests and quantitative tests. The qualitative tests are

1. Shake and bake tests
2. Torture tests
3. Elephant tests
4. Environmental Stress Screening and Burn-in

The main advantage of the qualitative tests is that one can increase the reliability by revealing and eliminating probable modes of failure. A good qualitative test is one which quickly reveals these failure modes that will occur during the life of the product under normal operating conditions. It is to be noted that the qualitative tests do not yield any data for the reliability or MTBF estimation of the product. In this section we describe briefly some of the tests carried out under Environmental Stress Screening.

The quantitative test is the Accelerated life testing. Unlike the qualitative tests described above the quantitative tests help in quantifying the life characteristics of the product and estimating the reliability of the product.

5.4.1 Environmental Stress Screening (ESS)

ESS can be defined as a process or a series of processes in which environmental stimuli such as rapid temperature cycling and random vibration are applied to electronic items in order to precipitate latent defects to early failures. Again, the purpose of any environmental stress screening (ESS) regimen is to expose the hidden defects that were introduced during the manufacturing process. More succinctly, manufacturing defects are precipitated from latent to patent. ESS, however, is not designed to find deficiencies in product design, although in many cases it does expose design deficiencies. Rather than design an ESS program to find design weaknesses, an important ingredient of product qualification must be the undertaking of environmental testing to ensure that the design is robust enough to meet its design goals.

In other words, ESS helps to convert latent defects of the products into detectable failures. ESS is designed to bring out defects introduced into the product by faulty components or the manufacturing process. These defects are often called latent defects, because they are not apparent unless the product is under the influence of some stress. Because ESS is an inspection step, it is not a value added process, and should be eliminated as soon as possible. It must be noted that ESS is not a simulation of the product's mission environment and has no relationship to the end use of the product. ESS is designed to apply appropriate stimulation (thermal, vibration etc) of high magnitude which will cause defective parts and workmanship errors to precipitate. It is also important to ensure that the applied stimulation does not approach the mechanical, electrical or thermal stress limits of any component to avoid accelerating the fatigue and/or causing damage. Each screening profile must be tailored (modified) for each module, unit, or assembly undergoing ESS. ESS is capable of producing valuable data for product improvement.

5.4.2 ESS in the infant mortality period

The underlying assumption for ESS is that there exist sub populations of latent defects, which, if not detected prior to shipment, would cause premature failures in service. These defects are responsible for the high initial failure rate of the bathtub curve, and if the ESS process is effective, they are precipitated to a large extent and the failure rate versus time curve is flat when the product is placed in service. Therefore, ESS can be used effectively only when the failure rate versus time curve has an initially decreasing slope. If this is not the case, there are no infant mortality defects, and any attempt at ESS will merely consume useful life without decreasing failure rate in the field. Some examples of latent defects that can potentially be converted to obvious defects are:

(a) Parts

1. Partial damage through electrical Overstress or ESD
2. Partial physical damage during handling
3. Material or process defects
4. Damage inflicted during soldering (excessive heat)

(b) Interconnections

1. Cold solder joint
2. Inadequate/ excessive solder
3. Broken wire strand
4. Loose screw terminations
5. Improper crimp
6. Loose conductive debris

Once the product passes through the early life stage it is usually reliable. The critical factor to control product reliability is to ensure that products have successfully passed through infancy before they are delivered to the customer. However no ESS program is perfect and some latent defects may remain in equipment after screening. MIL-HDBK-344 gives the following equation for determining maximum allowable number of latent defects:

$$FR \leq \left[\frac{1}{\text{Required MTBF}} - \frac{1}{\text{Inherent MTBF}}\right]\left[\frac{1}{\text{Safety Margin}}\right] \quad (5.25)$$

where:

FR is the Failure rate from latent defects
Required MTBF is the customer required value
Inherent MTBF is the predicted MTBF
Safety Margin typically ranges from 1.5 to 2

5.4.3 Burn-in

A special case of ESS is burn-in, which is the screening of components and assemblies at elevated temperatures, under bias, to precipitate defects prior to shipment. Therefore, burn-in should not be confused with more general ESS. However, sometimes, the term burn-in is used interchangeably with ESS.

As the electronics industry has matured, component technology and assembly techniques have changed profoundly. Products designed with vacuum tubes were the first to be subjected to environmental stress screening in the form of "burn-in". For these products, high temperature burn-in was the best stimulus for precipitating latent defects. Early transistor and integrated circuit technology defects were also effectively stimulated using high temperature burn-in. During the early days of electronics industry, most product defects were component-related defects. Today components have become so reliable that the failures are generally due to assembly and manufacturing processes. This has been supported by a study conducted by Motorola.

In 1990 Motorola conducted a study of burn-in effectiveness. They found that after burn-in only 0.000658% of the units failed an electrical test. Motorola's conclusion was that burn-in, prior to usage, does nothing to remove many failures *but may cause failures due to additional handling*.

5.4.4 ESS Tests

When performing ESS, it must be thought of as a process rather than a test. There are no accept/ reject criteria, and failures provide important information about product weaknesses. Failures are necessary to effectively predict a product's reliability. Some of the tests that are a part of the ESS process are described below.

- **Temperature cycling**: temperatures are varied between high and low values at a specific rate of change within a thermal cycling chamber. This process is repeated (cycled) at a specific time interval. The ability of the component or product to withstand temperature transitions is measured. Some of the failures that could be observed in temperature cycling are: solder joint failure, embrittlement of materials and crack. The corrective actions include use of better materials, components and improving the manufacturing process.
- **Vibration**: random and sinusoidal frequencies are implemented in the process to determine mechanical stress limitations. *Vibration test* is considered second highest precipitator of faults after *temperature cycling*. In Sinusoidal vibration a pure (ideally) sine wave excitation is given to the test item, whereas in random vibration, the excitation contains all frequencies in the given bandwidth. Random vibration is considered to be more effective because of the presence of all the frequencies in the given bandwidth. The principal effects of vibration are resonance damage and

fatigue damage. These can be limited by stiffening the mechanical structure and by controlling resonances.

- **Mechanical Shock** : Mechanical shock test is used to determine the ability of the product to withstand the shock that it might undergo during its transportation and/or service environment
- **Moisture/ Humidity tests**: Humidity/ Moisture tests are conducted during the development and evaluation or manufacture of equipment to ascertain its ability to resist the effects of humid atmosphere on the product. The humidity ranges from 5% to 98% and is generally combined with various temperatures to make the test more effective. The failures that could be addressed by this test are: loss of dielectric strength, corrosion and fungus growth. The corrective actions include applying adequate seals, moisture resistant parts and conformal coating.
- **Sand and Dust Test**: Sand and Dust is a test used during the development and evaluation of equipment to ascertain its ability to resist the effects of a dry fine sand laden atmosphere. This test simulates the effect of sharp-edged dust particles, which may penetrate into cracks, bearings, and joints, and cause a variety of damages such as fouling moving parts, making relays inoperative, forming electrically conductive bridges with resulting "shorts" and acting as a nucleus for the collection of water vapor, and hence a source of possible corrosion and malfunction of equipment. This test is applicable to all mechanical, electrical, electronic, electromechanical devices for which exposure to the effects of a dry duct (fine sand) laden atmosphere is anticipated, for example, all equipment meant to operate in a desert environment.
- **Salt and fog**: salt atmosphere conditions are simulated to evaluate the product's robustness in an ocean environment during storage, shipment or use. Exposure to salt fog is primarily results in corrosion of metals, although in some instances salt deposits may result in clogging or binding of moving parts and failure of electrical insulation. In order to accelerate this test and thereby reduce testing time, the specified concentration of moisture and salt is greater than is found in service. It is also useful for evaluating the uniformity (that is thickness and degree of porosity) of protective coatings.
- **Temperature/ humidity**: components or products are subjected to a specified temperature and relative humidity to understand the effects of these conditions. High/ low temperatures during storage/ operation: components or products are tested in storage or under operating conditions at high and low temperature extremes.
- **Rain**: wet conditions are simulated to determine the level of hermeticity that is sufficient to protect against rain penetration. The rain test is conducted to determine the effectiveness of protective covers or cases to

shield equipment from rain. This test is applicable to equipment that may be exposed to rain under service conditions.

- **Fungus**: The fungus test is conducted to determine the resistance of equipment to fungi. Fungi secrete enzymes that can destroy most organic substances and many of their derivatives. Typical materials that support fungi and are damaged are cotton, wood, linen leather, paper and cardboard. For fungus test, the test items are tested for several days (typically 28 days) at 30^0C and 95% humidity.
- **Solar radiation**: The sunshine test is conducted to determine the effect of solar radiation energy on equipment in the earth's atmosphere. For the purpose of this test, only the terrestrial portion of the solar spectrum is considered. The specified limits and energy levels provide the simulated effects of natural sunshine. The ultraviolet portion simulates natural sunshine in a general way, and is considered representative of irradiation in most geographical locations. Sunshine causes heating of equipment and photo-degradation, such as fading of fabric colours, cracking of paints, and deterioration of natural rubber and plastics. Many plastics tend to harden and lose their elastic properties on exposure to ultraviolet light. Sunshine tests are applicable to equipment that may be exposed to solar radiation during service or unsheltered storage at the earth's surface or in the lower atmosphere.
- **Electromagnetic radiation**: The equipments are tested for: 1. Generation of interference to other equipment/ objects, and 2. Susceptibility to interference from other equipment / sources. Corrective actions include grounding, shielding & filtering and power isolation & conditioning. Some of the examples of equipment that generate electromagnetic interference are printers, relays etc.

These are just a few of the ESS tests that are available to determine failure rates and service life in components/ products. The failure rates obtained during testing can be used to facilitate improvements in the design of a product or manufacturing process to meet customer expectations.

The ESS process is not complete without comparing the results to the standards and specifications that apply to the customer's needs. These criteria are applied to the product to identify the performance level that is expected of it within its intended environment. The product may be intended to fit the requirements of many different commercial, industrial, military, or international standards and specifications such as MIL-STD-810F and IPC-9701 along with additional customer specified criteria.

ESS programs which are applied during the development and production phases can yield significant improvements in field reliability and reduction in maintenance costs. Application during development can reap significant savings in test time and costs as a result of eliminating or reducing the number of latent defects prior to qualification tests. The benefits for the manufacturer include insight into the sources, better control of rework costs, decreased warranty costs and the opportunity to determine corrective actions that eliminate the sources of reliability problems from the product or process. A by-product of this activity is a more rugged product which may enjoy a higher demonstrated MTBF.

5.4.5 Quantitative accelerated life tests

Accelerated life testing can be grouped under two categories, namely usage rate acceleration and overstress acceleration. Usage rate acceleration can be analysed similar to that of life data analysis methods. The overstress acceleration method is the testing method in which the stress is increased in a stepwise manner. The rationale behind these tests is that if we wish to understand, measure and predict any event, we must observe the event!

5.4.5.1 Usage rate acceleration

Use rate acceleration is applicable for products which are not used continuously for a long time. For example, bread toaster, microwave oven, washing machine etc. In these cases one can accelerate the time by operating the toaster more frequently in a day to observe the failure. The most important thing to remember in these cases is that cycles of operation and cycling rate (or frequency) should not affect the cycles to failure distribution That is to say that reasonable cycling rate should be used to simulate actual use. The cycling frequency should be low enough to return the test unit to steady state after each cycle.

5.4.5.2 Overstress acceleration

Overstress acceleration is applicable for products which are in continuous use. Here the tests are planned in such a way that the product failure could be observed in the shortest possible time. This is achieved by applying stress (es) that exceed the stresses that a product will encounter under normal use conditions. The time to failure data obtained under these conditions are then used for extrapolating the usage conditions. The normal stresses used here are temperature, humidity, voltage, pressure, vibration and combinations of stresses to accelerate or stimulate the failure mechanisms. The stresses chosen should be such that they accelerate the failure modes under consideration but do not introduce failure modes that the product will not encounter in its use conditions. Thus the stresses should be such that they are outside the product specifications but inside the design limits.

Analysis of accelerated life test data then consists of determining an underlying life distribution that describes the product at different stress levels and a stress-life relationship that quantifies the manner in which the life distribution changes across different stress levels. The most commonly used life distributions include the Weibull, the exponential, and the lognormal. The exponential distribution can be used only if the underlying assumption of a constant failure rate is justified. From this distribution selection we determine the life characteristics. For example in the case of Weibull distribution two parameters β (shape parameter) and η (scale parameter) are the life characteristics. In the case of exponential distribution the parameter λ and the inverse $1/\lambda$ (the mean life) represent the life characteristics.

After obtaining the life characteristic parameters we need to select a model that describes a characteristic point or a life characteristic of the distribution from one stress level to another. Some of the stress life relationships are the well known Arrhenius relationship used for analysing temperature stress, Inverse Power law (used for stresses which are non-thermal in nature) and Temperature-Humidity relationship. These relationships are given below with respective parameters.

Arrhenius relationship

$$L(V) = C.e^{\frac{B}{V}} \tag{5.26}$$

Where:

- L represents a quantifiable life measure, such as mean life, characteristic life, median life
- V represents the stress level (in absolute units if it is temperature)
- C is a model parameter to be determined (C>0)
- B is another model parameter to be determined

Inverse Power Law Relationship

$$L(V) = \frac{1}{K.V^{n}} \tag{5.27}$$

Where

- L represents a quantifiable life measure, such as mean life, characteristic life
- V represents the stress level
- K is a model parameter to be determined
- n is another model parameter to be determined

Temperature-Humidity Relationship

$$L(U,V) = A.e^{\left(\frac{\phi}{V} \cdot \frac{b}{U}\right)} \tag{5.28}$$

Where:

- Φ , b and A are the three parameters to be determined
- U is the relative humidity
- V is temperature (in absolute units)

Once we have selected an underlying life distribution and the stress life model to the accelerated test data, the next step is to determine the parameters in the stress-life relationship model using any suitable method say graphical or least square method and maximum likelihood principle. After estimating the life distribution parameters and the stress life model parameters a variety of reliability information about the product can be derived such as:

- Warranty time
- The instantaneous failure rate, which indicates the number of failures occurring per unit time
- The mean life which provides a measure of the average time of operation to failure

5.4.6 Highly accelerated life testing (HALT)

The same principle of step-stress in the accelerated testing is extended to observe the failure in much shorter time. This principle was developed by an American engineer, Dr Gregg Hobbs and is fully described in [Hobbs(2001)]. Here no attempts are made to simulate the operational environment except possibly as the starting point for the step stress application. No limits are set on the stress levels and we apply whatever stress might cause failures to occur as soon as practicable, while the product is under continuous operation. Then we analyse the failures as described earlier and improve the design. By applying stresses well in excess of those that will be seen in service, failures are caused to occur much more quickly. The times or cycles to failures in HALT will be several orders of magnitude less than would be observed in Service. Thus we get the time compression in several orders of magnitude to observe the failures. Therefore we obtain time compression of the test programme by orders of magnitude, and much increased effectiveness. This generates proportional reductions in test programme cost and time to market, as well as greatly improves reliability and durability.

It is important to note that reliability/durability values cannot be demonstrated or measured using HALT. An accelerated stress test can provide such information only if the cause of failure is a single predominant mechanism of failure, so that we can correlate it with the type of stress applied and use appropriate relationships cited in the previous section. Generally in HALT we apply various combinations of stresses and it is difficult to derive stress-life relationships. In HALT we stimulate failures as quickly as possible using highly unrepresentative stresses, it is impossible to relate it to any reliability quantifiable such as MTBF. HALT approach can be applied to any kind of product technology. It can be applied to major systems, assembly level or components.

5.5 SAFETY AND HAZARD ANALYSIS

In order to assure reliability in the design from safety perspective certain probability of risk should be evaluated before the launch of the product. The traditional reliability assessments discussed so far does not suffice this requirement. The reliability assured is based on the mission in question and the mission time of the product. The customer would like to make sure that the design has catered for all the safety requirements. Thus certain qualitative analyses of reliability tools are carried out which finally give certain quantifiable analysis such as probability of risk. Some of these tools are the Failure Mode Effects Criticality Analysis (FMECA), Fault Tree Analysis (FTA) and Safety Hazard Analysis (SHA). We describe these tools in detail in the following sections.

5.5.1 Fault tree analysis (FTA)

FTA was first developed by HA Watson of BELL Telephone laboratories in 1961-62 during Air Force study contract for the Minute man Launch Control system. The first published papers were presented at the safety symposium sponsored by the University of Washington and the Boeing

Company. The Fault Tree Analysis also gained its importance from the nuclear industry following the Three Mile Island accidents in 1979. In 1981 US Nuclear Regulatory Commission (NRC) issued the Fault Tree Handbook NUREG-0492. FTA is extensively used where safety analysis is required and the systematic tree construction helps prevent oversights. Because of its logical, systematic and comprehensive approach it is capable of bringing out the design and operational weaknesses which might have escaped the design phase. This methodology helps as a good decision support tool in bringing out the design and operational weaknesses in complex systems and helps the managers and engineers to efficiently uncover and prioritise safety improvements. A major advantage of the FTA is its ability to address human errors, which the FMEA does not.

Fault tree analysis is considered to be a "Top-Down" approach by which each level of fault is expanded to its required input occurrence until a primary occurrence is defined. The undesired event is the top event and then attempt is made to find out all possible failures responsible for the top event. These causes are connected by the logical gates (such as 'OR', 'AND') to the top event.

5.5.1.1 Construction of FTA

The first step in the construction of a fault tree is the identification of the top event. Careful definition of top event is extremely important to the success of the fault tree analysis. The top event should be neither too general nor too specific. The event should be such that the event could have a critical impact upon safety of the general public, or operating and maintenance personnel.

All the causative failures of the top event are identified and connected to the top event through logical symbols (eg: OR, AND gates). This completes the first level of the tree. The process is repeated by considering each of the failures in the first level as a top event to obtain lower levels of the tree. A typical fault tree is composed of a number of symbols which are described in Table 5.1.

We can categorize them as primary event symbols, intermediate event symbols and Gate symbols. A typical fault tree can be seen in Example 5.4 (Figure 5.3). The primary events of a fault tree are those events, which for one reason or another have not been further developed. They mark the end of every branch in the tree. These are the events for which the probabilities will be provided using which we can calculate the probability of occurrence of the top event. Then we have the intermediate event symbols and logic gate symbols. Several examples can be seen in Henley and Kumamoto (1981) for use of all the gate and event symbols.

Using these symbols one can construct a fault tree for a chosen top event. A particular top event may be only one of many possible system hazards of interest. Large complex systems may have several hazards identified which mean a fault tree has to be constructed for each of those identified hazards. Once the FTA is constructed two types of analyses can be carried out: Quantitative and Qualitative.

Quantitative analysis

A quantitative analysis of a fault tree is finding the probability of occurrence of the top event. The probability of the top event is calculated by assigning probabilities to each of the basic events appearing in the fault tree. The probabilities of the basic events are generally given by the field experience. The probability of the top event is calculated applying two fundamental rules to AND and OR gates starting at the bottom of the tree.

Multiplication rule:

When an output results because of two or more input events occurring at the same time, as in an AND gate the output probability is computed by multiplying the input probabilities. This rule assumes that the input probabilities are independent of each other. For example, carburetor and the radiator failures in an automobile are independent of each other because if one fails, the other is not affected. If a component fails as a result of the failure of another component its probability of failure is not independent. However, most fault tree components are statistically independent of each other.

Addition rule:

For an OR gate, where the occurrence of any input causes the output to occur, the probability calculation is based on whether the events are mutually exclusive or not. If two events are mutually exclusive, such as high temperature and low temperature (if one occurs the other cannot occur at the same time), the probabilities are simply added. For mutually exclusive events:

$$P(\text{output}) = P(\text{event } 1) + P(\text{event } 2) + \ldots + P(\text{event } n)$$

If the events can occur at the same time (if the events are not mutually exclusive), such as high temp and high pressure then:

$$P(\text{output}) = 1-(1-P(\text{event } 1))\,(1-P(\text{event } 2))\ldots(1-P(\text{event } n))$$

Table 5-1. Symbols used in fault tree analysis

Event symbol	Event symbol name	Event description
Primary Events		
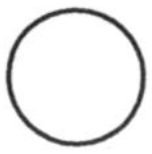	Basic event	A Basic initiating fault which requires no further development.
	Conditioning event	Specific conditions or restrictions that apply to any logic gate .Used with inhibit gate
	Undeveloped event	An event which is not further developed either because it is of insufficient consequence or because information is unavailable
	House event	An event which can be logically set to true or false depending on its application to the tree. Either occurring or not occurring
Intermediate event symbols		
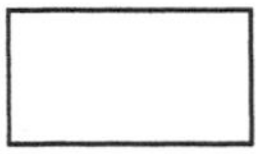	Intermediate event	A fault event that occurs because of one or more antecedent causes acting through logic gates. The top event of the tree is also represented by the same.
Gate symbols		
	AND gate	Output fault occurs if all the input faults occur.
	OR Gate	Output fault occurs if at least one of the input fault occurs
	Priority AND Gate	Output fault occurs if all of the input faults occur in a specific sequence
	Inhibit Gate	Output fault occur in a specific sequence (the sequence is represented by a conditioning event drawn to the right of the gate).

If the order of magnitude of probabilities is very small, say less than .001 then they can be added together like probabilities of mutually exclusive events. The answer will be very close to the exact value.

Qualitative analysis

The qualitative analysis of a fault tree includes definition of minimal cut sets of the tree. The minimal cut sets are obtained using Boolean reductions of the fault tree. The minimum cut sets give all the unique combinations of component failures that cause the system failure. The qualitative importance gives a qualitative ranking for each component with regard to its contribution to system failure. The common cause or common mode evaluations identify those minimum cut sets consisting of multiple components which because of common susceptibility, can all potentially fail due to a single failure cause.

A Cut set is defined as any basic event or combination of events whose occurrence will cause the top event to occur. Finding the cut sets for a given fault tree is a simple but repetitious task. The two simple rules to follow are:

- An 'AND' gate increases the size of a cut set.
- AN 'OR' gate increases the number of cut sets.

The minimal cut sets are obtained by eliminating the redundancy among the cut sets obtained. Thus a minimal cut set is defined as the smallest combination of events which, if they all occur will cause the top event to occur. By definition the minimal cut set is that combination of primary events which is sufficient to cause the top event to occur. If any one of the failures in the top event does not occur then the top event will not be caused by this particular combination of the minimal cut set.

The order of the cut set plays a very important role. The order is the number of events present in a cut set. Thus if there are too many of single order cut sets one has to be cautious of the top event in consideration. A single event or component that can cause the top event is a potential weakness in the system. Present day customers would like to know the risk in a product safety and specify the order of cut sets. The customer may not like single order cut set. Also the common mode failure of the cut sets is to be identified and remedy measures provided in the design for avoidance of the same.

The cut sets are also useful in bringing out the diagnostics provided in a design. Most products are designed today with hardware-software combinations. The fault codes are set through diagnostics and early warning for a non availability of a function in a system is intimated to the operator. For example, the present day automobiles are provided with various wheel-slip control functions by which the control of the vehicle is made safer through automatic interventions for various road conditions. With the diagnostics available for such functions the driver is intimated of the non availability of the function and he can maneuver the vehicle without depending on the function which is announced to be non-available.

A major advantage of the FTA is its ability to address human errors which the FMECA cannot address. FTA is particularly useful in safety analysis where the block diagramming discipline helps prevent oversights. Safety of the general public or operating and maintenance personnel could be studied. A small number of clearly differentiated top events can be explicitly defined. FTA is preferred where there is a high potential for human error and software error contributions. FTA helps in simplifying maintenance and trouble shooting.

EXAMPLE 5.4

The construction and the analysis of a fault tree are illustrated using an example of a breath analyzer. The breath analyzer is basically a screening instrument which is used for instant testing of the breath of a suspected drunk person. Air blown into the instrument flows through a sensor which is made of tin oxide heated to a particular temperature. Alcohol in the breath lowers the resistance of the sensor element by reducing tin oxide to metallic tin in a reversible reaction. The fall in resistance is measured to indicate the percentage of alcohol contained in the air breathed which in turn is related to the blood alcohol of the person breathing out.

The sensor is a commercial unit of high reliability. Absorption of vapors in ambient air over long periods of non-use can create a condition of low resistance which may take several heating cycles to cure. For this reason, the Breath Analyzer heats the sensor to a higher temperature than the operating value, temporarily to release absorbed gases. A precise voltage controlled and programmed power supply as well as measuring circuitry comprise the complete unit. Indication of critical alcohol content is through a flashing LED and an audio beep, both coming on simultaneously.

Table 5-2. Cut sets for the breath analyzer

S. No.	Gate no	Event Name	Probability	Effective probability
1	G1.1.1	Total failure of the sensor	0.0001	0.0001
2	G1.1.2.1	Sensor got stuck in high alcohol position	0.01	0.01
3	G1.1.2.2	Sensor got stuck in low alcohol position	0.001	0.001
4	G1.2	Air passage blocked	0.02	0.02
5	G1.3.1	Critical components in the PCB, (Stabilized Power failure)	0.003	0.003
6	G1.3.2	Failure of power supply to the PCB	0.01	0.01
7	G1.4.1 G1.4.2	Battery voltage low Diagnostic for battery fails	0.001 0.0005	0.0000005
8	G1.5.1 G1.5.2	LED indicator failed Audio indication-beeper failed	0.01 0.05	0.0005

Top-event: The breath analyzer not screening the alcohol content of a breath is the most critical event and is chosen as the top event. Figure 5.3 shows the construction of the tree for the event. From the tree constructed the cut sets have been derived. Table 5.2 gives the cut sets and the respective probability of each basic event in the tree. We find that for this example we have 6 single order cut sets and two cut sets of order 2. Using these probabilities and the rules stated above the probability of the top event is calculated. We find that the probability of occurrence of the top event is 0.044601. Suppose the system is operated for say 100 hrs we can presume about 4.5 failures or say 5 failures in the duration. The probability figures used in the example are purely from the point of view of an academic exercise.

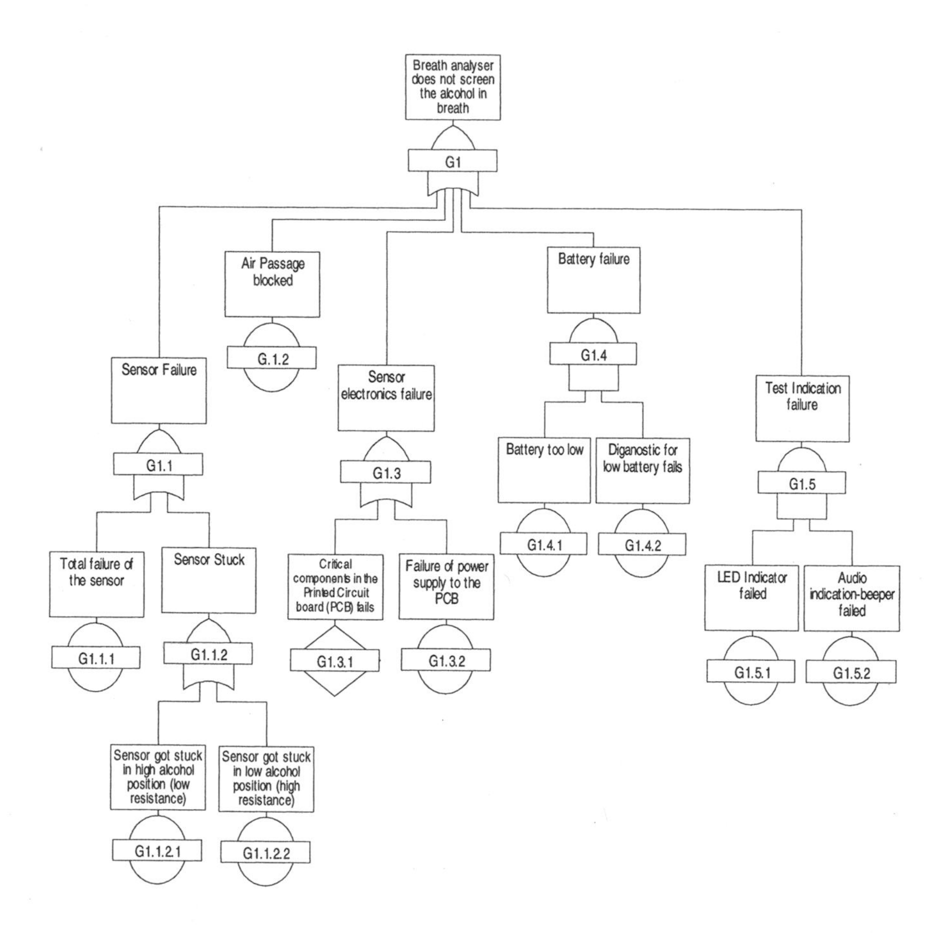

Figure 5.3. Fault tree of a breath analyzer

5.6 FAILURE MODE EFFECTS AND CRITICALITY ANALYSIS

Failure Mode Effects and Criticality Analysis (FMECA) is one of the effective reliability analysis methods. In this analysis, various failure modes of the individual component/subsystems are considered and their effect on the overall system performance. FMECA could be described as organization of knowledge about potential failures. This task is a design aid for assurance purposes. The analysis is often called bottom-up approach in contrast to the top-down approach of FTA. The results of FMECA serves as the main input to define Built in Tests (BIT) that can automatically detect and isolate system failures as they occur and increase system availability. FMECA also serves as a dictionary of failure modes for safety and logistics analysis. The use of FMECA is generally limited by the time and resources available and the capability to derive a sufficiently detailed database at the time of the analysis (for example, accurate system definition, the components present in the system/subsystem, up-to-date drawings, failure rate and hazard rate and so on).

FMECA is helpful in studying the different modes of failure for a component and its effect on the overall system performance. The analysis comprises two parts namely quantitative and qualitative. The acronym FMECA can be split into two FMEA and CA:

- Failure Mode Effects Analysis (FMEA)
- Criticality Analysis (CA)

FMEA is a qualitative analysis of the failure modes whereas CA is a quantitative analysis. In criticality analysis each failure mode is quantified by assigning a criticality number which is a product of its base failure rate, failure effect probability, failure mode ratio and the operational time.

5.6.1 Failure mode effects analysis (FMEA)

The purpose of the FMEA is to study the effects of item failure on system operation and to classify each potential failure according to its severity. In FMEA the component identification with its function, functional aspect in the mission of the system, different modes of failure, any compensating provision, and their causes are listed. The effects of each failure mode on the system performance have been divided into four severity categories. The MIL-STD-1629A gives the guidelines to carry out the FMECA analysis. The severity classifications for each failure mode as per MIL-STD-1629A are:

Severity I - **Catastrophic** - A failure which may cause death or weapon system loss (i.e Aircraft, tank, missile, ship etc.).

Severity II - **Critical** - A failure which may cause severe injury, major property damage, or major system damage and results in mission loss.

Severity III - **Marginal** - A failure which may cause minor injury, minor property damage, or minor system damage and results in delay or loss of availability or mission degradation.

Severity IV - **Minor** - A failure not serious enough to cause injury, property damage, or system damage, but results in unscheduled maintenance or repair.

The typical columns that would appear in a FMEA sheet can be seen in the case study given at the end of the chapter (Table 5.13). This format is referred to as TASK 101.

5.6.2 Criticality analysis

The purpose of the criticality analysis (CA) is to rank each potential failure mode identified in the FMEA Task 101, according to the combined influence of severity classification and its probability of occurrence based upon the best available data. The documentation of CA is accomplished as per the standard format given in Table 5.15. This format is referred to as Task 102 which supplements Task 101 and shall not be carried out without completing Task 101. The criticality number for each failure mode is calculated by taking the product of failure rate, failure effect probability, failure mode ratio and time of operation.

Failure probability/failure rate data source: When the failure modes are assessed in terms of probability of occurrence, the failure probability of occurrence level shall be listed. This is required when failure rate data are not available from prediction. When failure rate data are available then the data source could be listed.

Failure effects probability: The β values are the conditional probability that the failure effect will result in the identified criticality classification, given that the failure mode occurs. The β values represent the analyst's judgment as to the conditional probability of loss being occurred and should be quantified as stated in Table 5.3.

Table 5-3. Failure effect probability table

Failure effect	β value
Actual loss	1.00
Probable loss	> 0.10 to <1.00
Possible loss	> 0 to = 0.10
No effect	0

Failure mode ratio α : The fraction of the part failure rate λ_p related to the particular failure mode under consideration shall be evaluated by the analyst and recorded. The failure mode ratio is the probability expressed as a decimal fraction that the part or item will fail in the identified mode. If all potential failure modes of a particular part or item are listed, the sum of the alpha values for that part or item will equal one.

Part failure rate λ_p: The part failure rate is either an estimated value or calculated through reliability prediction.

Operating time : The operating time in hours or the number of operating cycles of the item per mission shall be derived from the system definition .

Failure mode criticality number (c_m): c_m is the portion of the criticality number for the item due to one of its failure modes under a particular severity classification. This criticality number gives a comparison between all failure modes for a particular item and helps to identify the most critical failure mode. Using the failure mode criticality number we can calculate the item criticality number by summing up the failure mode criticality numbers. This is useful in identifying the most critical item in a system.

In criticality analysis if the numerical value of failure data is not available then a qualitative approach could be made. Simple qualitative probability of occurrence levels may be defined as follows (as per MIL-STD-1629A):

1. Level A - Frequent: $0.20 < p < 1.0$
2. Level B - Reasonably probable: $0.10 < p < 0.20$
3. Level C - Occasional: $0.01 < p < 0.10$
4. Level D - Remote: $0.001 < p < 0.01$
5. Level E - Extremely unlikely: $0.0 < p < 0.001$

Where p is the failure probability of occurrence during item operating time interval, defined as a single failure mode probability of occurrence. This value of p could be supported by any experimental data; else it becomes subjective.

5.6.3 Risk analysis through FMEA

The failure modes of a component can be identified and prioritized by calculating Risk Priority Number (RPN). This is more a qualitative approach and commonly used in industry to assess the severity of the failure. For each mode of failure we can assign the factors in the scale of 1 to 10 for likelihood of occurrence, severity and detectability. Tables 5.4, 5.5 and 5.6 give some guidelines for assigning these ratings. The tables can be modified with more categorization with manufacture's/user's experience about the failures in the system. However increasing the number of categorization will be more difficult to manage with the subjectivity involved in this analysis. The RPN can range from 1 to 1000. The high values of RPN should be designed out by using better components or introducing better detectability. The probability of occurrence can be reduced by introducing redundancy if possible. This subjective method can be used in the early phase of the design as an iterative process before any test data are available. The advantage of this method is that there is no debate on the numerical values, but rather a comparative technique, in which one can compare the failure modes with the previous result.

Table 5-4. Categories of likelihood occurrence

Criteria	Rating	Possible Failure rate
Remote probability of occurrence. It would be very unlikely for these failures to be observed even once	1	0
Low probability. Likely to occur once, but unlikely to occur more frequently	2 3	1 : 20000 1 : 10000
Moderate probability. Likely to occur more than once	4 5 6	1 :2000 1 :1000 1 : 200
High Probability. Near certain to occur at least once	7 8	1:100 1:20
Very high probability. Near Certain to occur several times	9 10	1 :10 1 : 2

*Table 5-5.*Categorization of severity

Criteria		Rating
Minor	A failure that has no effect on the system performance. The operator would probably not notice.	1
Low	A failure that would cause slight annoyance to the operator, but that would cause no deterioration to the system.	2 3
Moderate	A failure that would cause a high degree of operator dissatisfaction (i.e high pedal effort, or radio buzz) or that causes noticeable, but slight deterioration in system performance.	4 5 6
High	A failure that causes significant deterioration in system performance, but that does not affect safety.	7 8
Very High	A failure that would seriously effect the ability to complete the task or could cause damage, serious injury or death	9 10

Table 5-6. Categorization of detectability

Criteria	Rating	Probability (%)
Remote probability that the failure remains undetected. Such a defect would almost certainly be detected during inspection or test	1	86-100
Low probability that the defect remains undetected	2 3	76-85 66-75
Moderate probability that the defect remains undetected	4 5 6	56-65 46-55 36-45
High Probability that the defect remains undetected	7 8	26-35 16-25
Very high probability that the defect remains undetected until the system performance degrades to the extent that the task will not be completed	9 10	6-15 0-5

5.6.4 Functional FMEA and Hardware FMEA

A FMEA analysis may be based on hardware or a functional approach. In hardware FMEA various failure modes of the hardware are considered. The functional FMEA is an alternative to a lengthy piece-part level analysis. The functional approach is used when hardware is not fully defined or if the system itself is very complex. In this approach functional failures are considered with a pointer to hardware.

In the hardware approach actual hardware failures are considered at the desired indenture level. Generally the indenture level is chosen based on the maintenance/repair policy such as line replaceable units (units which are replaced at line maintenance) to start with. For example, for an electronic design the indenture level can be either at the PCB (Printed Circuit Board) level or at the components present on the PCB. If we are replacing the PCB at the repair level it is better we do the FMEA for a PCB. Depending on the cost of replacement and the criticality of the PCB one can think of carrying out the component level FMEA for the critical components in the PCB.

In a mechanical design the choice of level of analysis to be performed is very important. The failure rate data is generally not available and we restrict our analysis to the qualitative FMEA. But when we want to decide the indenture level in a mechanical design, the restriction comes with the resources available. Considering the fact that different hardware are obtained from various vendors, the designer may not be able to break those subsystems into components for carrying out the FMEA at the component level. In such cases it is often better to produce a short FMEA, which considers the failure modes at sub assembly level with clearly defined failure effects rather than an extremely lengthy component analysis. Thus the functional FMEA is useful when the hardware items cannot be uniquely identified in a subsystem or in early design stages when hardware is not fully defined. In this approach function failures are considered (e.g no display on the console, memory lost). It is to be noted that a functional failure mode can become a hardware failure effect in a hardware approach FMEA. The functional FMEA is identical to the normal FMEA except that failure modes are expressed as 'Failure to perform particular sub-system functions'. The functional FMEA should consider both primary functions (that is functions for which the subsystem was provided) and secondary functions (that is, functions which are merely a consequence of the sub-system's presence). The case study at the end of the chapter has an illustration of this FMEA. (Table 5.11)

5.6.5 Advantages of FMECA

FMECA is useful in analyzing the various failure mechanisms and mitigating the failure modes. From FMECA, the critical components can be identified and listed separately. Failure modes that can be eliminated at very small cost can be promptly eliminated. The analysis is also helpful in determining the operational and maintenance requirements and in developing maintenance manuals.

5.6.6 FMECA and FTA in Design

As a part of reliability analysis FMECA (Failure Mode Effects and Criticality Analysis) and FTA (Fault Tree Analysis) are carried out for the system. FMECA is a preliminary design evaluation procedure to identify design weaknesses that may result in safety hazards or mission abort situations and also to assess their severity and adequacy of fault detection. Fault Tree Analysis on the other hand, provides an objective basis for analyzing the system failure and its causes, identification of critical components by connecting the component failures logically and systematically. FMECA is often referred to as inductive logic whereas the type of logic used in FTA is deductive. The hazard analysis involves a mixture of inductive and deductive logic [Benjamin 1990].

FMECA and FTA are essentially complementary to each other. FMECA is a bottom up approach and FTA is a top down approach. In general the main difference between FMECA and FTA is that FTA addresses human errors whereas FMECA does not. The functional FMEA described earlier certainly has an analogy with the FTA. In FTA we think of the hazards in the system level where as in functional FMEA we try to narrow it down to subsystem functions. Both FMECA and FTA can be performed without any duplication of work and we can link them together. The FTA could be developed down to the level of component failure and these basic events can be linked to the corresponding component FMEA. Thus we will have a comprehensive failure analysis of the system using both FTA and FMEA.

5.7 HAZARD ANALYSIS

The hazard analysis (or the safety analysis) refers to process or equipment failure or operability. The purpose of the analysis is to determine safety parameters, which would help prevent the possibility of damage to the system and/or personnel. The safety analysis serves as an aid to establish design criteria and as an evaluation tool for the subsequent assessment of design for safety. The goal is not merely a good design; rather the goal is a successful design which anticipates problems in advance, implements corrective action on time and maximizes the operability of the system while minimizing life cycle costs. Often tests cannot be carried out under conditions that duplicate the actual field environments, which can cause human errors, such as wrong wiring, poor soldering and welding. The safety analysis is carried out to consider the effect of the equipment failure on the system and welfare of humans.

A detailed preliminary hazard analysis on a system will help us document the list of top events which will be the necessary candidates for FTA. The consequences of the hazard could be analyzed for all possible situations of the system in its operation. Based on the consequences the mishap potential could be worked out and we can group the effects into critical, severe, moderate and minor. This will help us identify those FTAs that need deeper attention and analysis.

5.7.1 Hazard analysis methodology

Hazard analysis is carried out for each equipment/unit as soon as the system architecture is formulated. The hazard for a system could be caused by any of the following conditions/stages:

1. The system is in operation
2. The system is idle
3. The system or a component of the system operates above or below the specified conditions
4. Hazards due to environmental interface
5. Hazards due to human errors
6. Storage condition
7. Handling and transportation stages
8. Testing, repair and maintenance action

A system has to be analyzed for the possible hazards which could be caused due to any of the above listed conditions at the earliest possible phase of the design. Then the hazard control measures could be implemented during preliminary design phase itself. The following criteria could be applied to control the hazard:

1. Design the hazard out of the product. If the hazard cannot be eliminated, try to minimize the residual risk
2. Design for a fail safe default mode by incorporating safety devices or fault tolerant features
3. Provide early warning through measuring devices, software or other means. The warning should be clear and should attract the attention of the concerned operator

4. Implement special procedures, training and drills

The format for carrying out hazard analysis is taken from MIL-STD-882C. The information compiled can be seen in the example 5.5 (Table 5.10). Every hazard has a description and a cause for it. But for the hazard to happen there should be a triggering event. For example, fire can break out if there is inflammable media around. But for the fire to happen some ignition source should be available like an electric spark or a lighted match stick. Thus when we describe the hazard we have to think of all possible causes and triggering events. The hazard is evaluated on the basis of its severity and its probability of occurrence. Tables 5.7, 5.8 and 5.9 give the guidelines from MIL STD 882C.

Table 5-7. Hazard severity categories

Description	Category	Definition
Catastrophic	I	Death, system loss or severe environmental damage
Critical	II	Severe injury, severe occupational illness, or minor system or environmental damage
Marginal	III	Minor injury, minor occupational illness, or minor system or environmental damage
Negligible	IV	Less than minor injury, occupational illness, or less than minor system or environmental damage.

Table 5-8. Probability Classification for Hazards as per MIL-STD-882C

Description	Level	Specific individual item	Fleet or Inventory
Frequent	A	Likely to occur frequently	Continuously experienced
Probable	B	Will occur several times in the life of an item	Will occur frequently
Occasional	C	Likely to occur some time in the life of an item	Will occur several times
Remote	D	Unlikely but possible to occur in the life of an item	Unlikely but can reasonably be expected to occur
Improbable	E	So unlikely, it can be assumed occurrence may not be experienced	Unlikely to occur, but possible

We take the combination of hazard severity and its probability of occurrence and form the hazard risk index. Then we follow the guidelines from the MIL-STD-882C given in the Table 5.9 for accepting or rejecting a hazard.

Table 5-9. Hazard priority guidelines from MIL-STD-882C

Hazard Risk Index (HI)	Suggested criteria
IA, IB, IC, IIA, IIB, IIIA	Hazard unacceptable
ID, IIC, IID, IIIB, IIIC	Hazard undesirable (higher management decision is required)
IE, IIE, IIID, IIIE, IVA, IVB	Acceptable with review by management
IVC, IVD, IVE	Acceptable without review

The corrective action is determined by the criticality rating. For category I and II hazards the mitigation is accomplished by the design. Inspection, testing, warning and procedural actions are not acceptable unless one is willing to accept the risk.

EXAMPLE 5.5

An example of airborne early warning system with some of its critical subsystems is given for understanding the various information we compile and the remedial measures taken for avoidance of the hazard. The subsystems considered are auxiliary power generating system (APGS) and hydraulic drive system. The hazard due to environmental condition is also described considering the effect of electromagnetic radiation to personnel working in the design or maintenance phase of the system. The analysis is described in Table 5.10.

1. The total power supply for the rotodome drive in the airborne early warning system is provided by the APGS. Thus APGS is a very critical subsystem for the successful mission of early warning system. In APGS if the turbine blade rotates at a high RPM then there is a chance of its getting detached and hitting the control surfaces of the aircraft which in turn would cause an accident resulting in a catastrophe. This has been taken care of by providing a containment ring over the turbine. In case the turbine fails due to high rpm this ring would hold it in its place and the flight will not be disturbed. The table indicates the hazard index as ID.

2. In the hydraulic drive system due to high temperature the hydraulic fluid may catch fire and the aircraft landing gears and brake may fail. This has been taken care of by providing a thermal switch which operates at 60°C. This cuts off the hydraulic power to the drive system in case of temperature rise. For this also the hazard index is ID. From Table 5.10 we see that the hazard index for both these hazards is **ID**. That is **I** stands for the hazard severity indicating a system loss [Table 5.7]. **D** refers to the probability of occurrence which tells us that the event is quite unlikely but possible to occur during the life of an item [Table 5.8]. As per priority guide lines this hazard is an undesirable one [Table 5.9]. However these hazards have been countered by providing proper safety measures.
3. Table 5.10 in the example also describes the effect of electro magnetic radiation that will be created due to continuous rotation of the antenna. While testing the system the person working nearby will be exposed to this radiation. As per international standards if the person is exposed to a radiation greater than 10 Kilo volts / meter for more than two hours he is susceptible for retina damage and skin problems. The long term adverse effects include increased heart rate, cancer and neurobehavioral changes. The hazard index for this hazard is **IIC**. As per priority guidelines given in Table 5.9 this is also an undesirable hazard. This can be prevented by defining the boundary for the safe area beyond which the person should take enough precautions such as having a proper metallic shielding to avoid the radiation effect.

5.8 FAILURE REPORTING ANALYSIS AND CORRECTIVE ACTION SYSTEM (FRACAS)

So far we have seen how analyses like FMECA, FTA, HA can be carried out and how they can be used during the design phase. However these analyses are based on the previous experience with the design and similar products. Often we do come across several failures during the test and installation phase. To over come such failures observed during operation in the true environmental condition in the future design, it is necessary to compile the list of possible failures in a systematic and structured way. The failures form the wealth of data in designing better product. Questions such as what to improve? and what to design against? could be answered with this wealth of failure data.

A comprehensive failure reporting analysis and corrective action program is extremely important to the achievement of reliability in equipment. It is necessary to have a detailed program which assigns responsibilities to appropriate personnel to carry out this documentation of failure data. Early elimination of the causes of failure is a major contributor to reliability growth and attaining customer satisfaction after delivery of the product. It is easier to implement the corrective action if the failure causes are identified in the preliminary design stage. As the design matures the corrective actions become comparatively more difficult to implement. Cost and time play a major role in each corrective action. Therefore, it is important to employ FRACAS early in the development phase.

The documentation should provide information on

1. What failed?
2. How did it fail?
3. Why did it fail?
4. How can such failures be eliminated in the future?
5. The corrective action suggested
6. The feedback of the corrective action taken

A typical FRACAS report would consist of the following steps:

1. Observation of failure
2. Details of the failure along with the conditions which existed during the failure
3. Confirmation of the validity of failure
4. Isolation and localization of the failure to the lowest level such that the defective replaceable component is identified
5. Replace the defective part and verify the corrective action by testing the system
6. Confirmation that the suspect item is defective
7. Failure analysis of the defective item
8. Correlation of the defect with the previous history available if any
9. Establishment of the root cause of the failure
10. Design team to determine if any design change to be done with respect to the corrective action
11. Incorporation of the recommended corrective action in the reliability development system
12. Continuation of the reliability development test
13. Establishment of the effectiveness of the proposed corrective action
14. Incorporation of effective corrective action in the production equipments

The accurate input data assures an effective FRACAS. MIL-STD-2055 gives more detailed information on the format and various steps involved in carrying out a FRACAS.

5.9 CASE STUDY – AIRCRAFT EMERGENCY STOPPING SYSTEM

The aircraft emergency stopping system is used to bring the aircraft to a halt on runway in case of aborted take-off or landing emergencies. The system aids the aircraft to halt its forward momentum with minimum damage to aircraft or injury to pilot. The major subsystems of the system are:

- Engagement System
  - Multiple element net assembly
  - Shear pin
  - Shear off coupling
  - Purchase tape
- Engagement System support
  - Suspension cable
  - Suspension cable anchor
  - Chain and shackle
  - Net anchors
  - Ferrule
- Stanchion System
  - Stanchion frame assembly
  - Base frame assembly
  - Motor and winch assembly
- Tape connector
- Energy absorber
- Pressure roller assembly
- Sheave assembly
- Tape retrieval system
- Electrical control system
  - Main control panel
  - Remote control panel
- Foundation

When an emergency landing is requested, a command is issued from the air traffic control tower. This command is received through the remote control located in the traffic control tower and the multiple element net assembly is raised with the help of stanchion system. As the net envelopes the aircraft, the pull exerted on the net releases the net anchors and breaks the shear pins in the shear off couplings, releasing the net top from the suspension system. The purchase tapes attached to the net end loops are pulled through the fair lead tubes and begin paying off the two energy absorber tape drums, thereby turning the rotary hydraulic brakes. This action generates a uniform braking force which smoothly decelerates the aircraft to a safe stop.

After the aircraft has been safely halted, the net is manually disconnected from the purchase tapes, removed from the aircraft and transported to the hanger for inspection and repair. The purchase tapes are rewound onto the energy absorber tape drums by the tape retrieval system. During tape retrieval the pressure roller system ensures tight wrap onto the tape drum. After the tapes are retrieved a new net is quickly installed across the runway thus allowing the system to be ready for another emergency stop.

5.9.1 Functional FMEA

The total system is mechanical except for a little electronics in the electronic control part. The functional FMEA for the system can be carried out in sequence of its functions. The functions required to stop the aircraft are as follows:

- Receiving the command from the Air Traffic Control
- stanchion system raise to raise the multiple net assembly
- Disconnecting the net assembly from its support
- Energy absorption by the energy absorber

The functional FMEA can be carried out for each of these functions and we can have a detailed FMEA and criticality analysis for the hardware component which plays a critical role in the particular function. As an example the second function is taken, the stanchion raise which is an important function once the command is given. Table 5.11 gives the functional FMEA carried out for the stanchion raise of the aircraft emergency stopping system. The functional FMEA carried out for all the functions identifies certain major subsystems and assemblies responsible for the particular function. The list is given in Table 5.12. Thus we can segregate the hardware subsystems playing critical role in the failure of the particular function. The FMEA in Table 5.11 has considered two major functions, stanchion not raising at all and stanchion not raising to the required height. Among these two functions the first one is more severe when compared to the second one. The subsystem/components responsible for the function can be taken on priority to conduct the hardware FMEA for them.

5.9.2 Hardware FMEA

The hardware component level FMEA was carried out for the total system. Table 5.13 gives one of the FMEA carried out for the components of stanchion frame. For illustrative purpose one FMEA is given. The FMEA carried out for all the important components of the system could identify components whose failures lead to severity I and severity II which are most important. Severity I leads to catastrophic failure and Severity II leads to mission failure.

Table 5.14 gives the list of components with Severity I and Severity II. This serves as a major aid in paying special attention on those during maintenance. A daily inspection is necessary on such components. Also during the design phase importance can be given towards their reliability and maintainability to assure higher availability of the system.

Table 5.15 describes the criticality analysis carried out for the stanchion system. For all the failure modes described in the FMEA (Table 5.13) the criticality numbers are calculated. The item criticality number is obtained by taking the sum of all the criticality numbers of the failure modes of that particular item.

Table 5-10. Hazard analysis of airborne early warning system

System: Airborne Early Warning System Subsystem:						Date: Sheet:		Analyst: Approved by:			
S.No	Equipment Name	Hazard Description	Causes	Accident Trigger Event	consequences	Impact on interfacing systems	Hazard Severity	Hazard probability	Hazard Risk Index	Safety provision made in existing design	Remarks
1	Auxiliary Power Generating System	Turbine Failure	Turbine RPM too high	Defect in rotor blade due to material defect	Engine power may not be available	Aircraft Hydraulic system environmental control system may get damaged	I	D	ID	Containment ring is provided over the turbine. This takes care of turbine not getting detached	Better material and design of blades to avoid the failure

System: Airborne Early Warning System Subsystem:						Date: Sheet:		Analyst: Approved by:			
2	Hydraulic drive system	Temperatu-re increase of fluid >70 deg C	Contin-uous rotation of rotodo-mes	Insufficie nt cooling system or failure of heat exchanger	Hydraulic fluid catching fire	Aircraft landing systems and brakes failure	I	D	ID	Thermal switch provided, which cuts off the hydraulic power to the drive system when temp > 60 deg C	
3	Airborne early warning system	Electro Magnetic Radiation	Continu ous rotation of rotodo-mes	Continuo-us exposure	Damage to retina and skin problems	Long term adverse effects on other organs of the body	II	C	IIC	Shielding has to be provided and barrier for safe area to be determined	The long term adverse effects include cancer and neurobeha vioral problems

Table 5-11. Functional FMEA of Stanchion System of Aircraft Emergency Stopping System

System Name: Aircraft Emergency Stopping System Subsystem Name: Stanchion System Reference Drawing Number:				Date: Complied by: Reviewed by:		
Function	Function Failure	Failure mode	Local effect	System Effect	Severity Class	Remarks
Stanchion raise	Stanchion Fails to raise	Geared brake motor of stanchion system does not function	The motor will not rotate to raise the stanchion	The stanchion not raised for aircraft stopping	I	
		Winch drum assembly fails	Winch cable will not wind properly on the drum	The necessary drive to raise the stanchion is not transmitted and the stanchion not raised	I	
		Improper splicing of	Stanchion frame loses the	The engagement becomes ineffective and	I	

System Name: Aircraft Emergency Stopping System Subsystem Name: Stanchion System Reference Drawing Number:				Date: Complied by: Reviewed by:		
		ferrule of lift cable	support	the a/c is not arrested		
		Weldment broken due to bending	Stanchion system itself broken	No engagement of the aircraft	I	
	Stanchion not raised to the required height (ht)	Weldment frame bent due to lifting or jerking while lowering	The top mount assembly of the stanchion is not supported fully	Stanchion height reduces and hence the height of the multiple net assembly reduces	III	Degraded mission depending on the ht.
		Pin in the top fitting assembly breaks	Extended height cannot be provided	Net may sag and centre ht of the net not taken care and the aircraft canopy may get hit by top horizontal straps and it may cause injury to the pilot	II	

Table 5-12. Subsystems function leading to severity I

Sl No	Function	Subsystem failure	System effect
1	Stanchion raise	Geared brake motor	Stanchion not raised for aircraft stopping
		Winch drum assembly	Stanchion not raised
		Lift cable	Engagement is ineffective
2.	Disconnecting net assembly from its support	Shear pin does not break	Ineffective engagement
3.	Energy absorption	Rotor vanes	The energy of the aircraft not absorbed and hence speed of the aircraft not reduced
		Brake force to the shaft is not transmitted	Rotor will not rotate and hence no energy absorption
		Purchase tape	Link between net assembly and energy absorber cut off and hence no energy absorption

Table 5-13. Hardware FMEA of Stanchion System of Aircraft emergency stopping system

System Name: Stanchion system Subassembly name: Reference Drawing No: Mission					Date Sheet: Compiled by: Reviewed by:						
I.D No.	Item identification	Function	Failure Modes & Causes	Mission Phase/Operational mode	Failure Effects			Failure Detection Method	Compen-sating provision	Severity class	Remarks
					Local Effects	Next Higher level	End Effects				
	Weldment frame	To support the top mount assembly	Weldment bending due to lifting, Engagement, Welding Limit switch failure	To support engagement system	Height of the stanchion system will reduce	Net Centre height reduces	Engage-ment is not proper	Inspection	NIL	III	In the long run stanchion will not be ready for use
			Weldment breaks due to bending or higher strength of shear pin		Stanchion system breaks	Needs replace-ment for next engagement	Engage-ment is not proper	Inspection	NIL	II	Cause injury to pilot – This is serious when one side breaks
			welding of few		Bending of weldment	Height of the stanchion	Net centre ht. Reduces	Inspection	NIL	III	Proper maintenance will take care of this failure

					Failure Effects						
			members break		frame	system will reduce					
	Link Cable bracket	Link between shock absorber and stanchion frame assembly	Welding breaks	To sustain shock after fully lowering	No connection with shock absorber	Jerk will come on weldment frame while lowering	Weldment frame will break	Inspection	NIL	IV	
	Lift joint assembly	To transfer necessary load to lift stanchion frame assembly	Lift bracket is broken	To support the Stanchion frame through lift cable	Stanchion will not be raised	No engagement	Aircraft will crash	Inspection	NIL	I	Daily inspection mandatory
	Hinge joint assembly	To support and allow the movement of stanchion frame assembly	Tearing off due to – twisting -weak welding	To raise the stanchion	Support of the stanchion lost	Damage to stanchion system	Improper engagement	Inspection	NIL	III	

Table 5-14. Components leading to Severity class I

Sl.No	Subsystem	Component	Failure mode
1.	Stanchion system	Lift joint assembly Lift cable	Lift bracket breaks Improper splicing of ferrule
2.	Motor and winch	Geared brake motor	Failure of brakes Mounting bolts not in position
3.	Winch and drum assembly	Lift cable	Improper splicing
4.	Purchase tape		Stitches in purchase tape off
5.	Tape connector	Shaft	Bending of shaft
6.	Engagement support	Ferrule Chain and shackle Anchor bracket	Improper splicing Links in chain break Weak welding
7.	Electrical control	Main control panel Remote control panel	Failure of critical component

Table 5-15. Criticality Analysis of Stanchion System of Aircraft emergency stopping system

System Name: Stanchion system — Date
Subassembly name: — Sheet:
Reference Drawing No: — Compiled by:
Mission — Reviewed by:

I.D No	Item identify-cation	Function	Failure Modes& Causes	Mission Phase/ Operational mode	Seve-rity	Failure rate data source	Failure effect Probab-ility	Failure Mode ratio	Failure rate	Operating time	Failure mode Criticality	Item criticality	Remarks
	Weldment frame	To support the top mount assembly	To support the top mount assembly	To support engagement system	III	NPRD-95	0.1	0.4	0.0005	15.0	0.003		
					II	NPRD-95	0.75	0.3	0.0005	15.0	0.00169		
					III	NPRD-95	0.1	0.3	0.0005	15.0	0.0023	0.00221	
	Link Cable bracket	Link between shock absorber and	Link between shock absorber and	To sustain shock after fully lowering	IV	NPRD-95	0.0	1.0	0.3327	15.0	0.0	0.0	

System Name: Stanchion system
Subassembly name:
Reference Drawing No:
Mission

Date
Sheet:
Compiled by:
Reviewed by:

		stanchion frame assembly	stanchion frame assembly										
	Lift joint assembly	transfer necessary load to lift stanchion frame assembly	transfer necessary load to lift stanchion frame assembly	To support the stanchion frame through lift cable	I	NPRD-95	1.0	1.0	0.3327	15.0	4.9905	4.9995	
	Hinge joint assembly	To support and allow the movem-ent of stanchion frame assembly	To support and allow the movement of stanchion frame assembly	To raise the stanchion	III	NPRD-95	0.1	1.0	0.0005	15.0	0.00075	0.00075	

Chapter 6

INSERVICE RELIABILITY

Science never solves a problem without creating ten more.
George Bernard Shaw

6.1 INTRODUCTION

An airplane has been described as 4 million parts flying in close harmony. Each of these components can generally fail in anyone of several ways. A simple nut may come undone because it was not tightened properly or it was not "locked" correctly or it was over-tightened and its thread was stripped or it was put on the bolt crookedly causing it to cross thread. It could have been the wrong size nut for the bolt or the wrong type of thread, the wrong type of material or the nut and bolt were too small or weak for the task. In each case, when the two parts being held together exert undue strain on the nut, it will not be able to withstand the stress and come off. The time when this happens may be the first time the aircraft takes off or it could happen after several successful flights. The time it lasts may be affected by the number of take-offs (and landings which should be the same number), the actual time in the air (flying hours), the number of times the door is opened and closed, the number of times the engine fairing is opened and closed and, if it survives long enough, the amount of corrosion it may have undergone.

In some cases, these failure modes will be competing, whilst in others they will be mutually exclusive. Unless the designer or maintenance manual has failed to specify a locking action (such as using a locking washer, a lock nut, or a split pin) then failure due to an unsecured nut coming loose is not likely to affect the whole fleet. Similarly, fitting a nut made from the wrong material or that is too large is also unlikely to affect the whole fleet unless these too are incorrectly specified in the design or maintenance manuals.

A pilot very nearly lost his life in a BAC 1-11 as it climbed through 17000 feet and the cockpit window in front of him parted company with the aircraft. On this occasion it was not due to incorrect nuts, rather 78 of the 81 screws used to replace the window during its last maintenance were of an incorrect diameter and length and were unable to withstand the build up of pressure within the cockpit. The error was traced back to the mechanic.

For a lowly component such as a nut, it is unlikely that its age is accurately known at the time of failure, particularly if it is part of a component which is regularly removed. With aircraft, it is quite a common practice to treat such items as "consumables" that is ones which are replaced every time the [parent] component is recovered or maintained. For many parts, especially those which are rather more expensive, replacement may only be made if the component is found to be in an unsatisfactory condition when inspected. In some special cases in which a failure of a component is both known to be age-related and can cause serious or even catastrophic consequences, replacement may be at or before the component reaches a certain age – generally referred to as life-limited parts.

As condition monitoring systems improve, components are becoming increasingly more likely to be removed before they actually fail. Methods such as performance trending and oil debris and vibration monitoring can often pick up deterioration in components without the need to take systems apart to carryout detailed inspections. Unfortunately, from a statistical viewpoint, this makes time-to-failure analysis less certain since it is generally not possible to determine when the component would have failed had it remained in service.

By contrast, if components are failing before they are expected to, particularly if these cause system unavailability or worse, safety concerns, considerable effort will be expended to isolate the cause. Once this has been found, it is very likely the part or parts implicated will be re-designed to what is generally referred to as a new modification standard (mod std). This action also causes problems with analyses since all of the data relating to the first mod std is generally of no relevance to the new one. Quite often components will be modified to meet new demands from the operators such as increased thrust, lower noise and nitrous oxide emissions, better fuel consumption and many more reasons. Although these may not be reliability related, they none the less are likely to change the reliability of both the parts changed and very often some that are not changed. In case of gas turbine engines, for example, changing some of the components is likely to change the frequencies of the vibrations induced. This could result in components spending more, or indeed less, time in the critical range of their natural frequencies and hence either fail more quickly or last longer.

6.2 DEFINITION OF INSERVICE RELIABILITY

In-service reliability is the probability that a given system or component thereof will remain in a state of functioning for some time t given it is used and maintained in a similar way to those that have preceded it. It is determined statistically from analysis of in-service usage data. It differs from any engineering prediction in so far as the analysis is based on observations. It is therefore a function of the accuracy and relevance of this data.

It is an unfortunate fact of life that there is currently no way of determining what the in-service reliability of a component or system will be, based on the design/engineering drawings. Reliability predictions are generally little more than "Engineering Judgment" which Richard Feynman suggested was nothing more than a guess (in his report on the Challenger Space Shuttle accident).

An Aloha Airlines Boeing 737 suffered from metal fatigue causing a large section of the roof of the fuselage to tear away, the same problem that caused several disasters with the first jet airliner, the de Havilland Comet in the early 1960's. Although by the time B737 incident occurred, a great deal more was known about metal fatigue and measures had been in place for some years to reduce the risks of such an incident, the calculations on this occasion had gone wrong. The aircraft had actually not flown a great many hours. Indeed many B737's had, at that time had flown far more hours than this one. What had led to its demise, however, was that it had flown a very large number of flights, all of relatively short duration mainly island hopping in the Hawaii archipelago where the typical stage length was some 20 minutes. The primary cause of metal fatigue in the fuselage is the stresses induced by changing atmospheric pressures rather than the length of time it spends at any given pressure.

Within gas turbine engines, the primary cause of failure of many components is more likely to be due to the accumulation of stress usually counted in "cycles". A [stress] "cycle" is the stress induced by taking the component from zero stress (idle) to maximum stress (maximum revs) and back to idle. In a commercial airliner, this corresponds very nearly to a normal flight or stage length. In combat aircraft, it is very much dependent on the mission profile and the way the pilot uses the controls; every movement of the throttles causes the rotational speed of the engine to increase or decrease adding to the stresses induced. In case of turbine blades centripetal forces are only one of the factors as the blades are also subjected to very high temperatures and vibrations causing thermal fatigue, metal fatigue and creep. The actual cause of failure of such a blade could be any one of maybe seven different and competing modes. Although not very often, the primary cause, corrosion can create "stress raisers" from which cracking can start which, if left undetected, is likely to lead to final failure.

Few years ago, a South American Air Force suffered the failure of a number of LP Compressor blades (the big fan blades at the front of the engine). This was followed by a similar problem faced by a southern European Air Force with blades that were older than those in the other fleet. Curiously no problems had ever been encountered with these blades with engines used by the Royal Air Force (RAF) that had been in service for a great deal longer than those of either of the other two fleets. After a great deal of analysis and investigation, it was discovered that the root cause of the problem was corrosion. Pitting of the aluminum alloy in one specific area was enough to cause a stress raiser which eventually resulted in the catastrophic failure of the blade. Apparently the RAF blades were routinely subjected to coatings of oil-rich smoke during aerobatic displays and this had reduced the risk of corrosion. By contrast, the engines in South America were regularly washed on Friday afternoons in pond water and left to dry over the weekend before being flown again on Mondays.

Software failures are notoriously difficult to predict (see also Chapter 8). Very often the cause of the failure is due to either a part of the program being activated in a way that had not previously occurred or data has been entered incorrectly. A common problem with Monte Carlo models (i.e. ones which use pseudo-random numbers) occurs when a random number very close to zero or one is sampled and gets rounded down to zero or up to one. If it is used with certain types of probability distributions it may be necessary to take the log of it or its complement resulting in a failure. Obviously, this type of error can be very easily avoided by simply checking the value before putting it into such a function or by using double precision which avoids the rounding.

Consider the maintenance support and development of a FORTRAN event based simulation model. Typically, this would be run at night on the IBM mainframe computer with each of the 10 passes taking up to an hour. Every now and again, an execution error would occur (by sod's law, in one of the later passes) which proved particularly difficult to isolate. The model was written with time-activated print switches that the user could enable by setting the [simulation] time of when they were to be switched on and off. It was necessary to do this because otherwise the amount of output would destroy several rain forests and take many years to trawl through. Unfortunately, activating these print switches, would cause the time set (which was stored in the form of a binary tree) to have extra elements in it which generally meant that some of the events would be activated in a different order than previously. This would then cause the random numbers generated to be used in a different way and hence cause divergence from the original run invariably causing the run to either fail at a different time (just before or after the printing was turned on or off!) or, sometimes it would not fail at all. The problem eventually was solved by rewriting the model in a different language (to a slightly different specification).

6.3 INSERVICE RELIABILITY DATA

At the operational unit (OU) level, e.g. aircraft, ship, car or wind turbine, we may know the number of operating hours (across a fleet of these OU or for a single OU) and the corresponding number of failures. In Figure 6.1, we can see the times between failures for 3 different systems. In each case the total operating time is 10 and the number of failures is also 10 thus the Mean [Operating] Time Between Failures (MTBF) is 1 for each case.

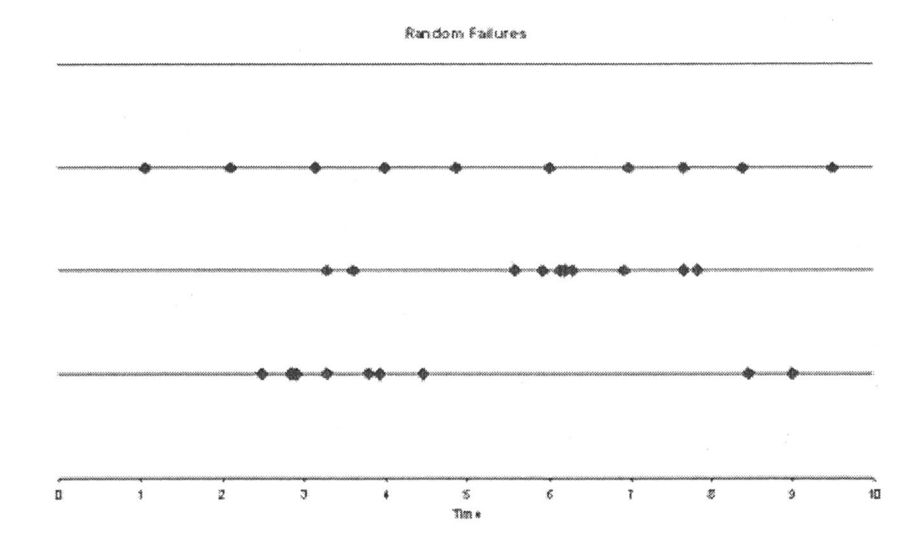

Figure 6-1. Random failures

Although the MTBF's for each of these three cases are the same, the pattern of failures appears to be quite different. We know that there are essentially three types of failure caused by poor quality control, accidental damage and wear. Typically, these can be described by Weibull distributions with different shape parameters (<1, = 1, >1) respectively. If we knew the actual ages of the components at the time of failure for each of the failures then we could perform standard parameter estimation methods to get estimates of the Weibull shape and scale parameters but before we do that, we should first try to determine the causes of each of the failures.

In the first example in this chapter, the case of the BAC 1-11, the cause was found by the Air Accident Investigation Board (AAIB) to be maintenance induced. The mechanic had simply fitted the wrong sized screws. In this case, the time to failure was less than 1 hour, just long enough for the airliner to reach 17,000 feet where the in-balance between the cabin pressure and the much lower atmospheric pressure was too great for the screws to hold the window in place. It could also be argued that it was a design fault as well since had the window been designed to be fitted from the inside to "plug" the hole in the fuselage the internal pressure would have simply forced it into this gap creating a tighter seal/fit.

By contrast, the second example – the Aloha B737 failure – was due primarily to excess stress caused by usage. It is what is generally referred to as an age-related failure. In this case, the aging unit was stress causing metal fatigue due to the number of times the cabin/fuselage had been pressurized and depressurized and hence to the number of flights (since every flight causes the cabin to be pressurized once and depressurized once).

After the loss of the Challenger Space Shuttle an investigation was carried out to determine the cause of the loss and ascertain whether it could have been prevented with a view to ensuring that a similar fate would not befall future missions. The investigation concluded that the failure had little to do with the age of the craft or its usage but was almost certainly due to the ambient conditions, in particular, the temperature shortly before take-off.

Most systems consist of a very large numbers of components. It can be proved mathematically that for such a system the times between failures will become exponentially distributed as the [operating] time increases indefinitely (Drenick, 1960). It is a well known fact that if the failures are exponentially distributed then there is no point in performing preventive maintenance and that any maintenance on the system will effectively restores it to a same-as-new condition.

Whilst this may be true, at the component level, most mechanical systems exhibit age-related failures caused by usage – wear or erosion – or natural aging such as corrosion. In practice, many failures due to these age-related causes can be and are avoided by judicious use of preventative and opportunistic maintenance. The use of condition monitoring can also reduce the number of actual failures. It should be recognized that although these practices can reduce the probability and hence the number of failures, they will not reduce the number of component replacements unless the failure of one component has a significant probability of causing the failure/replacement of another – often referred to as secondary or caused damage. Inside a gas turbine engine there are very few gaps so if a blade or other piece of material becomes detached there is a very high probability that it will impact at least one other component as it passes back through the engine and finally out into the exhaust stream. Given the high velocities and energy levels of these detached pieces, and the lack of a free passage it is almost inevitable that they will cause damage as they make their way out of the engine. The failure of a bearing may either cause the shaft/axle to seize or could allow it too much freedom to move from its axis of rotation and hence cause damage to other parts. If the cam-belt in a piston engine car breaks there is likely to be damage to the camshaft, pistons, valves and possibly the cylinder sleeves or linings which is, of course, why it is recommended to have this item replaced after a given mileage to reduce the risk of an actual failure and hence its associated damage.

In many systems it is not always possible to determine the precise cause of failure. In gas turbines, there is a measure known as the TGT (turbine gas temperature) margin. As the components in the engine, in particular, the compressor blade tips wear, the amount of cold air reaching the combustion chamber is reduced and hence output thrust decreases for a given level of fuel-burn. To achieve the required thrust on take-off and through the climb, it is therefore necessary to increase the amount of fuel entering the engine which means that the engine will be running hotter. If the exhaust gases are hotter then the temperature of the turbine blades will also be higher then the [TGT] margin will be reduced. These blades are kept from melting by having cold air blowing through them as their melting point is very much lower than the exhaust gas temperature. When the margin falls below a given value, the engine has to either be down-rated (and possibly moved onto a smaller aircraft) or the component(s) causing the problem have to be replaced. It is very rarely the direct cause of one component, indeed very often none of the components is badly worn or damaged rather it is a combination of several; as a result, it may not be possible to determine which one is the primary cause and which are secondary causes.

We can see much the same thing in a car engine. As the engine gets older so the components start to wear or deteriorate. One symptom of this is that the fuel consumption (miles per gallon, kilometers per litre) tends to increase. Another might be that the engine starts to burn more oil or the acceleration is not as good or it has difficulty getting up steep hills or it misfires or it simply fails the emission tests as part of its annual Ministry of Transport (MOT) test. Again, it is unlikely to be the result of a single component, rather it is the combined effect of several, some of which may be relatively cheap and easy to replace, others not so. The fact that we replace sparkplugs, [contact breaker] points, condenser, HT leads, oil, oil filter, air filter and reset the timing does not necessarily mean that anyone of these was the primary cause but it is a lot cheaper to do this than to replace the piston rings, big and little end bearings, camshaft, head gasket, inlet and exhaust valves, cylinder head and other parts of the engine that may be worn or indeed the whole engine.

The fact that we may have repaired or replaced some of the components in the system clearly does not mean the system is restored to an as-good-as-new condition. Replacing the tyres, brakes, clutch or even the engine does nothing to reduce the amount of corrosion in the bodywork or the wear in the gearbox. There are very few ten-year old cars that sell on the secondhand market for more than their original price.

With military equipment and in some cases where the operator owns a large fleet of similar operational units (OU) recovery of the OU may be achieved by removing a "line replaceable unit" (LRU) and replacing it with a spare of a similar type. The removed LRU may in turn be recovered by removing and replacing modules and the modules by removing and replacing or repairing parts. In each case the recovered component, whether it is an LRU, module or part, will be added to the stock of spares to be reused at some later date.

The main benefit of this approach is that the OU can be recovered often in a matter of minutes or a few hours whereas to repair or replace the parts may take weeks or even months. This means the operator needs fewer OU to meet a given level of availability (at the fleet level) although they will generally need more spares at the lower levels of indenture (LRU, modules and parts). From an in-service reliability viewpoint, this introduces additional problems; a component that started in a training aircraft based in Wales, could next find itself in combat in the desert, part of an aerobatic display team or even on-board an aircraft carrier, for example. Each of these different uses is likely to affect the amount of stress and hence rate of wear that many of the components will experience in any given mission. At the same time, even if the component stays with the same fleet, it is still necessary to track it through its life if we are to get reasonably accurate ages at the time of failure and/or replacement. With thousands of individual components in a system and maybe hundreds or even thousands of such systems belonging to the operator, keeping track of every instance is by no means easy.

If we consider a turbine blade, there could be 50 to 100 identical blades around the disc in the engine. Each of these blades is coated with a protective material. To inspect such a blade, it is usually necessary to first remove this coating, carry out the inspection then recoat it. It is normal practice to do batches of maybe several hundred at a time both to strip and to recoat. Although each blade will have its own serial number, there is actually very little space to put this on the blade. It cannot go on the aerofoil as this would interfere with the function and probably introduce a point of weakness. The only place is on the root which may have an area possibly 2-3 millimeters wide and no more than 10 mm long. If the serial number is put on before being coated then it will not be visible until it has been de-coated and if added after coating, it will be lost when the coating has been removed (or rubbed away). If the blades from a given set are always kept together then if we know the age of the set, in theory we know the ages of the blades, except, of course, if one, or more but less than the whole set, of blades are replaced during any given recovery. During the life of the set, all of the blades are likely to get replaced, some more than once so the age of the set will not necessarily be the same as the age of any given blade within it.

With a complex system, the number of failures, even for a large fleet, is unlikely to be sufficient to allow accurate estimates of the time-to-failure distribution parameters, at least not until the fleet has been in operation for many years. "Engineering judgment" is used before EIS (entry into service) to generate estimates of these parameters (or, to be precise, of one of these parameters – the mean or sometimes the B_{10} point also known as percentile life) but, for the vast majority of components, these estimates are based on precious little, if any, evidence. In many cases they are little more than a sort of ritualized allocation exercise. The Ministry of Defence (or equivalent) will specify what the "reliability" of the system has to be, typically quoted as a failure rate of x [failures] per 1000 hours. This is then divided amongst the main sub-systems (for an aircraft this would be avionics, airframe, propulsion, weapons and miscellaneous, say) in a fixed, but arbitrary way (probably based on the parts count). For the propulsion system, this value will be further divided between the accessories (pumps, control units, etc) and the engine or engines. The value allocated to each engine is then further divided between the modules and on down again eventually to the piece parts. Some parts in an engine, typically the discs and shafts, are safety critical (i.e. if they fail, there is an unacceptably high probability that the aircraft and/or lives will be lost). These are often referred to as Group A parts. If their times-to-failure are strongly age-related (as is often the case), they will be given a "predicted safe cyclic life", "life limit" or "hard life" usually determined via accelerated life testing (ALT) using test rigs and then calibrated as in-service data becomes available. This "life" is set such that there is an extremely low probability of failure occurring before this age. These parts can therefore be taken out of the [allocation] equation since they will be replaced during planned maintenance and should never fail. Some other parts may also be thought to have strongly age-related failures or might be capable of being continuously monitored. The former may be given soft lives such that they will not cause the system to be taken out of service but will be replaced if they have exceeded their soft life when the LRU containing them is subjected to invasive maintenance. This is unlikely to avoid all failures of these components but could reduce the numbers and hence allow the other parts to have higher allocated failure rates. Health and condition monitoring may also be used in a similar way.

When a system first enters into service, there will, of course, be no in-service reliability data, so we shall have to rely on estimates made based on engineering judgment and statistics derived from other similar systems operated in similar ways. What constitutes "similar" is very much up to the judgment of the Reliability Engineer. In the absence of, or possibly in addition to such data, the Reliability Engineer will need to consult the designers to find out what, in their opinions, are going to be the most likely causes of failure. Initially, it is recommended that this is restricted to asking three qualitative questions for each component: how likely is it to suffer inherent failures; extrinsic failures and usage-related failures. This can then be followed by further questions to determine whether all of the components are likely to be at equal risk or whether these types of failure will be restricted to small batches or even individuals.

Let us firstly look at inherent failures. Usually this type of failure is described by a Weibull distribution with a shape $\beta < 1$. An alternative way of looking at it, particularly in the case of quality or maintenance-induced failures is to consider how many instances of the component may be at risk and how often similar components have failed for one, or other, of these reasons. Suppose in a particular F1 Grand Prix season, there were 16 races with 24 cars starting on the grid in each race and suppose on 7 occasions one of the cars stalled on the grid. Assuming all cars/drivers are equally susceptible, then we might decide to model this using a binomial distribution with $n = 24$ (for a given race), $p = 7/384$ and $q = 377/384$. The mean is given by $np = 7/16$ and the variance by $npq=2639/6144$ from which we can predict that for any given race, there is a probability of nearly 36% that at least one driver will stall.

Suppose during the same season, almost 90% of cars (345) survived to the first tyre change and that of these 80% had a second tyre change and out of all of these events, just 2 wheels were lost on or before exiting the pit lane. Now, there were 24x16 - 7 = 377 starts. At the first pit stop, some 1380 wheels were changed and at the second 1104, so our sample size is 2484 with just 2 failures giving us a failure rate of 805 per million.

6.4 INSERVICE MTBF

For many years, the main "measure of reliability" has been the MTBF or its reciprocal, the constant failure rate (CFR) usually denoted by λ. In the case of the CFR, this is often expressed as "x failures per 1000 hours", if "hours" is the chosen measure of age.

There are several reasons why the MTBF has become the ubiquitous "measure of reliability" even though it patently does not satisfy the definition given earlier as neither it nor the CFR are probabilities. It is not clear who started this confusion but the US Department of Defense and the UK Ministry of Defence have certainly done much to contribute to it and perpetuate it by their references to it in such standards as MIL-STD-1388 and the UK equivalent Def Stan 00-60.

An advantage of a single parameter is that it is easy to refer to. It is also easy to calculate the maximum likelihood estimate $(1/\hat{\lambda})$ of the MTBF as:

$$(1/\hat{\lambda}) = \frac{1}{N}\sum_{i=1}^{n} t_i \tag{6.1}$$

Where N is the number of failures, n is the number of operational units and t_i is the current age of the ith operational unit. This can be simplified even further by noting that the summation gives the total operational time to date (summed over all of the operational units). Note that N can be greater than n, as it is possible for an operational unit to fail several times during its life. It also means that it is not necessary using this estimator to know the individual times to failure, and hence there is no need for parts life tracking (see later in this chapter). A problem with this can occur if any modifications have been made to the operational units which may affect their reliability. Then, using the cumulative operational time and number of failures means the full effect of these changes will not be seen in the estimated MTBF. (The mathematics supporting this is described in more detail later in this chapter).

Another, perhaps less obvious, advantage of this "measure" is that the user need know nothing about statistics or be concerned with complications such as variances, hypothesis testing and goodness-of-fit. It is probably fair to say that most practitioners concerned with "reliability" in fact have no formal education in statistics or probability theory and probably do not realize that what they are doing when estimating the MTBF has anything to do with statistics. There is certainly a great deal of confusion between "MTBF" and "constant failure rates" with many apparently believing that the times to failure are themselves constant.

For any distribution except the exponential, the conditional probability of failure is dependent on the age of the item. For most mechanical systems, the times to failure of many of the components will tend to increase with age – i.e. as the component gets older so the probability it will fail in the next increment of time will increase. Since, also in most of these systems, failed components may be replaced by new, reconditioned or repaired ones during the life of the system, the ages of the components can quite quickly get out of synchronization making it extremely difficult, mathematically, to predict the future behaviour of the system (in terms of when it will be in a state of functioning or a state of failure). This problem does not occur if it is assumed all TTFs (time to failures) can be modeled by the exponential distribution.

In addition, if the system reliability can be described as reliability block diagram in which all of the [critical] components are in series, then the CFR for the system is simply the sum of the CFR's of each of the components (on the diagram), as described in Section 4 of Chapter 4. The formula for components in parallel is slightly more complex but still very simple. If, however, the TTF's have increasing or decreasing hazard functions then these two formulae become complex multiple integrals which invariably are not closed and can usually only be evaluated (in a realistic time) by using Monte Carlo simulation.

Two theorems which have done much to promote the use of the exponential distribution are Drenick's (Drenick, 1960) and Palm's theorems (Palm, 1932). Drenick's Theorem, basically says that the hazard function of a system, under steady-state conditions, will become asymptotically constant (Kumar, et al, 2000). This means that whenever the system fails, it is instantly replaced with a new one which is identical in every way to the replaced one. It also means that the system will continue to be operated in exactly the same way. In practice, if the component, which caused the failure of the system is replaced with a new one to the same specification then eventually, the mean of the [operational] times between failures of the system will tend to a constant value – i.e. the MTBF.

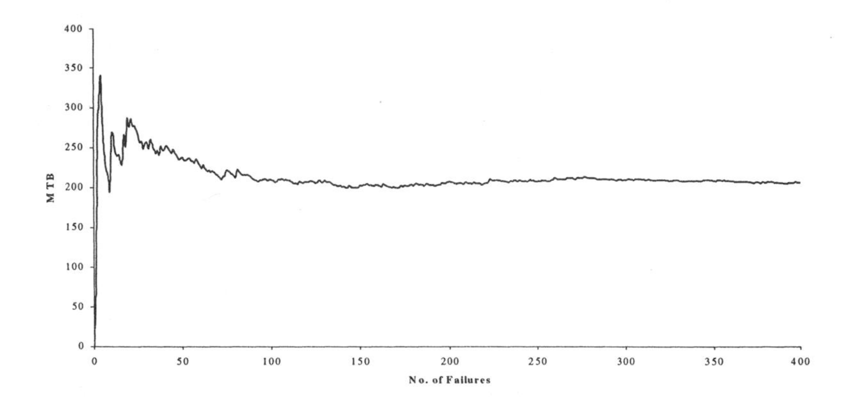

Figure 6-2. MTBF of a system

It should be noted that the assumptions which are required for Drenick's theorem to be valid are rarely satisfied and, even when they are, a great deal of useful information is lost in the process. Figures 6.2 and 6.3 show how Drenick's Theorem might work in practice.

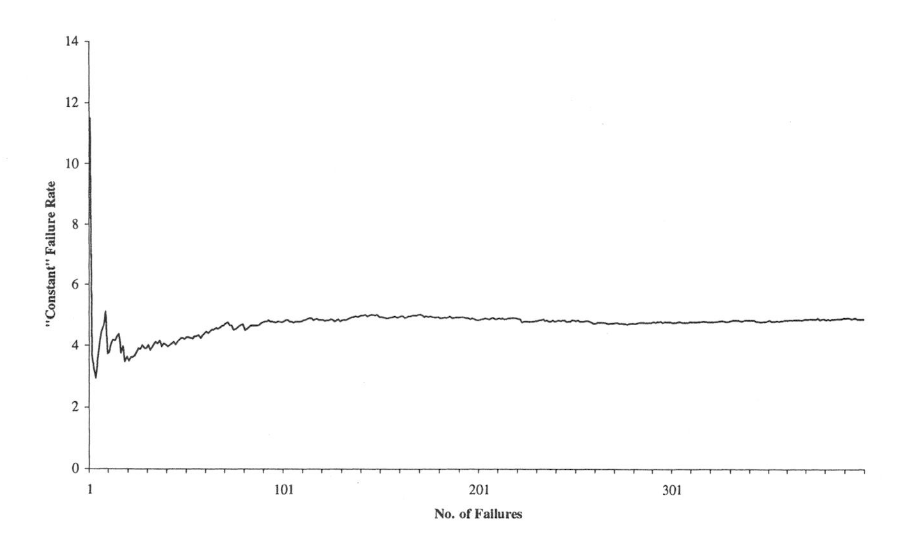

Figure 6-3. Constant failure rate of a system

In Figure (6.2), after around 100 failures of a given system, the time between failures has settled down to a fairly flat line at a little over 200 hours. Equivalently, the failure rate in Figure 6.3 tends towards a constant value at a little under 5 per 1000 hours (KHr). The first graph was produced by taking the cumulative age (i.e. time since new) and dividing this by the number of failures up to that time. The second gives the reciprocal of this "MTBF" multiplied by 1000 to give failure rate in failures per KHr. In both cases only the first 400 failures of the system were used (It should be noted that this is a fictitious system and that the times to failure were sampled from their respective Weibull distributions).

Here it can be seen that all that needs to be known is the age and the number of failures in order to calculate the MTBF. In this case, it was for a single, albeit multi-component, system but the same would have applied had the failures been for a number of identical systems. It can also be seen that *steady-state conditions* were not reached until the system was aged around 20,000 hours or equivalently, when it had suffered 100 failures.

If these times-to-failure (TTF) had happened to be for a gas-turbine engine in a combat aircraft, then 20,000 hours would represent approximately 50 years worth of flying. On the other hand, had they been from a commercial aircraft, this might have represented about 5 years, although it is unlikely to have remained in service for that long as this number of failures would simply make it uneconomical. Similarly, if this had represented a gas-turbine powered electrical generator, then 20,000 hours would have taken less than 2½ years of continuous operation although again, the operators would not have been very impressed with 100 failures. (Typically, both the commercial and power generation operators would be expecting no more than 1-2 failures in this length of operation).

In Figure 6.4 instead of a system consisting of just one operational unit (e.g. aircraft), there are now 100 "identical" operational units (OU) operating in parallel. The graph shows that as the fleet ages (uniformly across all OU) the "failure rate" increases, albeit at a decreasing rate so that by the time each OU has achieved 10,000 hours (a typical life for a military aircraft) the failure rate has climbed to 4.4 per KHr. This indicates that rather than achieving a "reliability growth" the system's reliability is in steady decline. Of course, this is what one would expect if no design changes were made to the operating units and, if one or more of their constituent parts suffered from [positively] age-related failures.

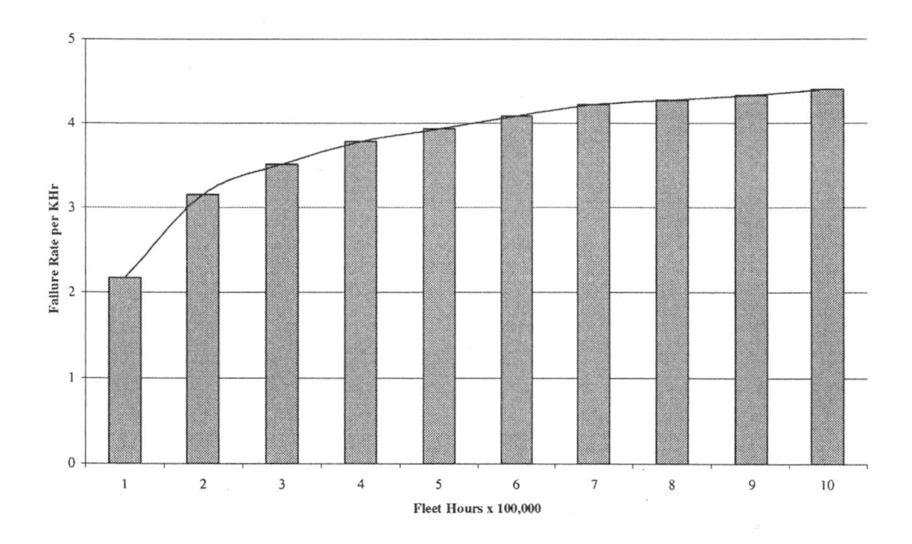

Figure 6-4. The failure rate for a fleet against the cumulative age of the fleet

For the exponential distribution, the probability density function (PDF) is given by:

$$f(t) = \lambda e^{-\lambda t} = \frac{1}{MTTF} e^{-\frac{t}{MTTF}}$$

Where $t > 0$, and λ is the failure rate and $(1/\lambda)$ is the mean time to failure (MTTF). Now, after a system has been operational for a period, the number of failures is (N_f) and a number of OU which are still functioning (N_s). Let T_i ($i = 1$ to N_f) be the ages of the OU at the time of failure and S_i ($i = 1$ to N_s) be the ages of the OU which are still functioning (i.e. that have not yet failed), usually referred to as *censored* or *suspended.*

The PDF (f(t)) gives the probability (or likelihood) that a failure will occur **at** time t_i, whilst $R(t_i)$ gives the probability (or likelihood) that a failure will **not** have occurred **by** time t_i. From this the *likelihood function (l)* can calculated as

$$l = \prod_{i=1}^{N_f} f(T_i) \prod_{i=1}^{N_s} R(S_i) = \prod_{i=1}^{N_f} \lambda e^{-\lambda T_i} \prod_{i=1}^{N_s} e^{-\lambda S_i} \qquad (6.2)$$

Taking logs (to the base e) gives

$$L = -\log(l) = -N_f \log(\lambda) + \lambda \sum_{i=1}^{N_f} T_i + \lambda \sum_{i=1}^{N_s} S_i \tag{6.3}$$

To find the value of λ which maximizes the likelihood, it is necessary to take partial derivatives with respect to λ and set this equal to zero:

$$L_{\max} = \frac{\partial L}{\partial \lambda} = 0 = -\frac{N_f}{\hat{\lambda}} + \sum_{i=1}^{N_f} T_i + \sum_{i=1}^{N_s} S_i \tag{6.4}$$

Rearranging gives

$$M\hat{T}TF = \frac{1}{\hat{\lambda}} = \frac{1}{N_f}\left[\sum_{i=1}^{N_f} T_i + \sum_{i=1}^{N_s} S_i\right] \tag{6.5}$$

To be pedantic, this should be differentiated again and checked that this is negative to be certain that this is a maximum. The maximum likelihood estimator (MLE) of the mean time to failure is simply the cumulative age of the fleet (i.e. total number of hours flown) divided by the number of failures (during that time). Thus the points plotted on the graph in Figure 6.4 are the "best" or maximum likelihood estimates of the mean time to failure for the operating units at various stages through their life, *assuming that the times-to-failure are exponentially distributed.*

In practice all OU are likely to be still functioning as they will have been recovered by the replacement of the failed components and returned into service. The S_i then are the ages (number of hours flown) since the given OU was last recovered. However, rather than using the times between failures, it is easier to use the current age of each OU and the number of failures each has experienced. Thus the formula becomes:

$$M\hat{T}TF = \frac{1}{\hat{\lambda}} = \frac{\sum_{i=1}^{N} A_i}{\sum_{i=1}^{N} F_i} \tag{6.6}$$

Where A_i is the age of OU i (from new), F_i is the number of failures of OU i (since new) and N is the number of OU.

Note that numerator in equation (6.6) is simply the total operational time for the system (fleet) and the denominator is simply the total number of failures to date. There is therefore no need to record the individual times-to-failure or, indeed, the individual ages of the OU, if the average daily, monthly, yearly usage of the fleet is known approximately.

It is because of this fact that the assertion was made earlier that few people who make these [MTBF] calculations know anything about statistics, probability theory, hypothesis testing or chi-squared goodness-of-fit tests. Given that the "analyst" never sees the actual times to failure or even the times between failures, there is really no way of knowing whether the times are constant, increasing or decreasing and hence will not be in a position to question the validity of the CFR assumption.

It will be noted, that if the hazard function is independent of the age, then the system can always be regarded as *as-good-as-new* no matter what has happened to it or how long it has been operating. In particular, every time the system undergoes corrective maintenance, it is restored to an as-good-as-new condition. It also means that because the probability of the system failing (in the next instance of time) is always the same; there is no benefit from replacing components before they have failed. In other words, preventive maintenance is pointless and repairing a component is just as effective as replacing it.

Returning to the example system used to generate the three graphs (Figures. 6.2-6.4), it is clear that this system is not in a steady-state condition and that there is at least one ageing process in operation. Unfortunately, without more information, there is very little that can be done to isolate this process or processes.

6.5 PARTS LIFE TRACKING

Figure 6.2 was obtained by taking the age of the system at the time of each failure. This is the same as recording a car's mileage every time [corrective] maintenance is performed upon it. Performing a Weibull analysis to fit the given times-to-failure (using median rank regression (MRR), see Chapter 7) yields a shape parameter (β) of 1.34 and a characteristic life (η) of 48,500 hours i.e. W[1.34, 48,500]. Linear regression gives a correlation coefficient of 0.98 which is not particularly impressive for 400 points. Plotting these points on [the equivalent of] Weibull paper yields the graph shown in Figure 6.5.

The shape of the graph strongly suggests that there are at least three failure modes present. The first part (bottom left) indicates the ages of the earliest failures are lower than would be expected (for a homogenous set with shape 1.34) suggesting some of the failures may be better modeled by a Weibull with a lower shape ($\beta < 1.34$). By contrast, the steep rise in the top right section suggests these were due to a failure mode better modeled by a higher shape ($\beta > 1.34$).

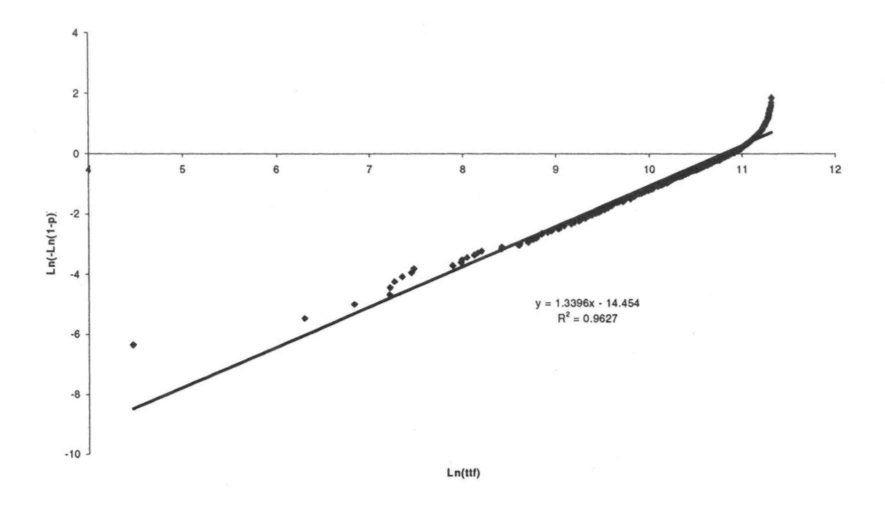

Figure 6-5. Weibull plot for system times-to-failure

The values of β and η can also be estimated by using the maximum likelihood method (see Chapter 7). Using Microsoft Solver, β is found to be 1.69 and the corresponding value of $\eta = 46{,}300$. Although the difference between 1.34 and 1.69 is not very large, the fact that the sample consists of 400 times to failure data strongly supports the suggestion that the sample is non-homogenous. Note that these results do not disprove Drenick's theorem, they merely confirm what we had already observed, that the system was not in a [mathematically] steady-state condition during this period.

Suppose the ages of the components were known at the times of failure along with the respective "primary" cause. "Primary" is used here to define the component which started the train of events which finally led to the failure. In a gas turbine engine, the failure of a turbine blade is, in itself, unlikely to cause the engine to fail. However, if it breaks off at its root, the resulting loose, but very high-energy piece of metal is likely to cause significant damage on its way out of the engine, usually sufficient to cause an engine failure.

Table 6-1. Estimates of the Weibull parameters

Component type	Actual		No. of failures	MRR		MLE		Bayes Scale
	Shape	Scale		Shape	Scale	Shape	Scale	
1	2	8400	11	1	8500	1.8	7400	6700
2	4	1000	94	3.8	1000	3.5	1000	1000
3	2	2200	42	1.8	2200	2	2200	2200
4	2	1900	51	2	1800	2.4	1800	1700
5	4	4200	22	2.1	4200	3.4	4000	4100
6	2	4500	19	1.9	4800	2.3	4700	4600
7	4	1800	50	2.9	1900	3	1800	1900
8	5	9800	8	5.2	10300	6.1	10200	10100
9	1	1000	91	0.8	900	0.9	900	900
10	2	7000	12	3.2	7200	3.7	7100	6700

Table 6.1 shows how well the MRR and MLE methods estimate the parameters of the Weibull distributions of the times to failure for each type of component. The final column also gives the best estimate of the scale parameter if the shape is "known" – often referred to as "Weibayes". There is a certain amount of empirical data to support the hypothesis that a given failure mode will have the same shape parameter in different systems (of a similar nature). If this is the case, then Bayesian Belief Nets can be used to estimate the shape from which it is then possible to estimate the scale parameter using the MLE formula.

The first observation to make from Table 6.1 is that all three methods give reasonable estimates of the scale parameter, even when the sample size is small (< 30). The MRR method also tends to give good estimates of the shape parameter in all but Component Type 5. For small samples, it has been recognized (Crocker 1996) that the MLE method tends to over-estimate the value of the shape parameter.

In the specific case of Component Type 5, Figure 6.6 illustrates the cause of the discrepancy. The two lowest aged failures appear to be too low. If these are ignored in making the estimate of the shape and scale, the revised estimates become 3.7 and 4100, respectively. There is, however, no reason for doing this although, it is likely that much effort would have been put into trying to find a difference between these and the remainder, i.e. to isolate a batch cause.

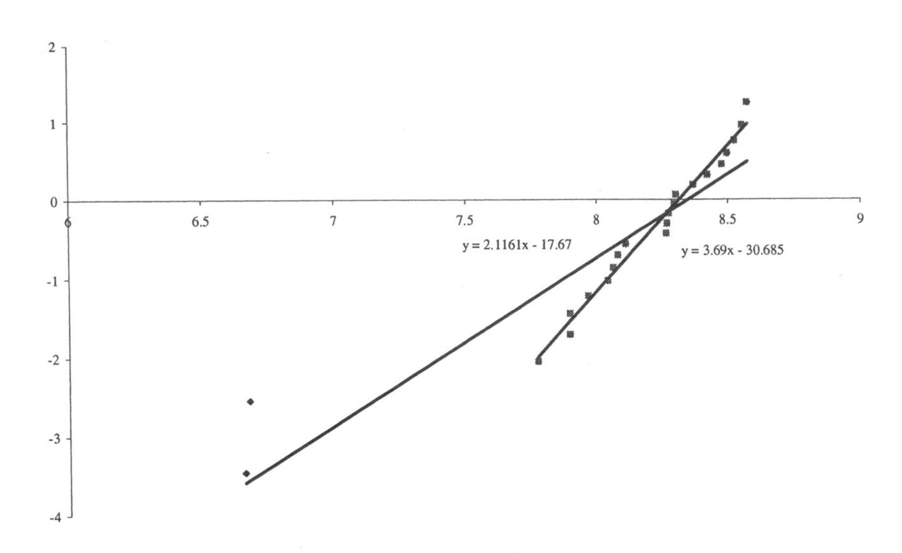

Figure 6-6. Weibull plot for Component Type 5

6.6 SAFETY CRITICAL PARTS

Not all parts are equally important, in other words, some parts are more important than others. In a gas turbine engine, for example, there is likely to be 20-30,000 individual items. Most of these are not critical, in so far as their individual failure is unlikely to cause an engine failure. However, there still remains a significant number that are critical. For some parts, their failure will have an unacceptably high probability of causing not just an engine failure but could result in the aircraft becoming uncontrollable. These parts are generally referred to as "safety-critical". Obviously, every attempt is made to minimize the number of these and to ensure that they have an extremely low probability of failing before the engine is decommissioned. Unfortunately, in some cases this is not possible and it then becomes necessary to take alternative action.

For many safety-critical parts, their times-to-failure are strongly age-related. This means that the hazard function is an increasing function of the age or, equivalently, the probability of failure (in a given period of operation, e.g. flight) increases as the age of the part increases. Note that the hazard function does not produce a probability as its value can exceed unity. For example, if the mean time to failure is 30 min then the hazard function will be 2 for an exponential distribution when the times to failure are quoted in hours, i.e. we would expect 2 failures per hour of operation.

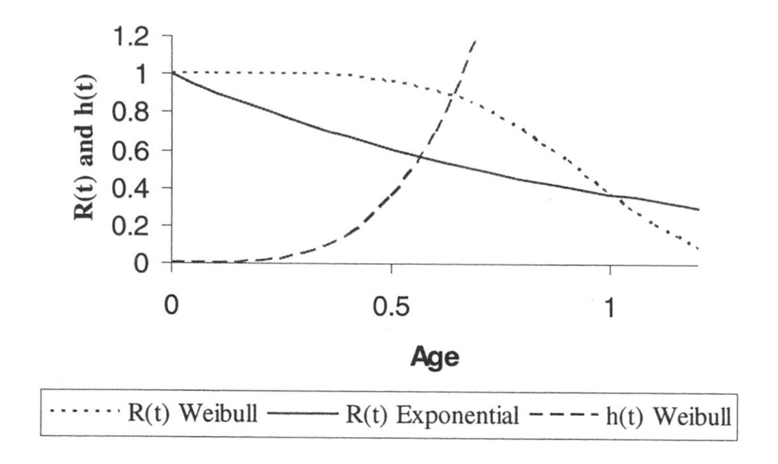

Figure 6-7. The reliability and hazard functions for a component whose times to failure are "strongly" age related.

In Figure 6.7, the hazard (h(t)) and reliability (R(t)) functions are plotted for a Weibull distribution whose shape (β) is 4.75 and characteristic life (scale, η) is unity (i.e. W[4.75, 1]). The reliability function is also plotted for an exponential distribution with the same characteristic life (i.e. W[1, 1]). These graphs show that there is a roughly 1 in 750 chance that the first component will survive for a time = 0.25 units whereas the second system could only be expected to survive a time = 0.0013 units. Thus if the units of time happened to be KHr, then the first would be "safe" for 250 hours whereas the second would only be safe for 1 hr 20 min. The paradox, however, is that the first would have to be replaced after 250 hours as the probability of failing in the next 250 hours would be much higher than 0.0013 (in fact 0.035) whilst the second could be flown for as many flights of duration of 1 hr 20 min as one wished with exactly the same probability of survival.

The unit of ageing is taken as a stress cycle. One stress cycle is defined as minimum stress to maximum stress and back to minimum stress. For a turbine disc, there is a further complication in that there is invariably a temperature gradient across the disc (from centre to circumference) and this too can vary according to the temperatures of the gases impinging upon the [circumference of the] disc and the cooling air which starts at the centre and radiates out across the disc (commonly referred to as *thermal transience*).

Before a disc's *release life* (or *hard life*) is set to the "safe life" a number of checks are performed. Typically, the initial release life will be set to half the safe life. As each of a certain size sample of discs reaches this release life, it will be removed from the engine and put back on test. If the mean number of stress cycles for this sample is not significantly less than the original [sample] mean, it is assumed that a "test stress cycle" is reasonably correlated with an "in-service stress cycle". The release life, in this case, may be increased to say ¾ of the "safe life" and the process repeated.

If the sample fails the test (or, indeed, if the original safe life is considered too low), the disc may be redesigned in order to strengthen it. If this increases its weight, it may then be necessary to increase the strength of the shaft which connects this turbine disc with its corresponding compressor drum as the extra weight will increase the torque on the shaft. A heavier disc will also require a greater force to turn it which means that the blades will have to extract more energy from the gas flow. This can either be done by changing their size/shape or by running the engine hotter (so there is more energy in the gas flow for the blades to extract). Increasing the size of the blades will mean it is very likely that the whole engine will have to be redesigned. On the other hand, running the engine hotter means the blades will need more cooling air to stop them from melting (as the gas temperature is generally considerably hotter than the melting point of the [turbine] blades). Diverting more air to cooling the blades means there is less to provide oxygen for the combustion process which means the gas flow will have less, not more, energy so it is back to the drawing board again. It is easy to see from this rather crude example that trying to improve the reliability after the engine has entered into service can be a costly process.

6.7 USAGE MONITORING

In the days before engine monitoring systems (EMS) it was impossible to know how long an engine had been run either "over-speed" or "over-temperature", usually as a result of too high a throttle setting for too long. If pilot reported that he had seen the temperature falling from within the "red zone", this could mean that the engine would need to be thoroughly inspected, possibly to the point of being stripped. It would not be until many days or weeks later when the engine and modules had been stripped and the components had been inspected that it could be determined whether any damage has incurred.

If your car is fitted with a tachometer or "rev counter", the chances are that it will have a section with a red background – usually from about 6,000 rpm upwards depending on the size of the engine and the type of car. If the engine is operated in this zone for too long at a time, it is likely to overheat causing the lubricating oils to breakdown and/or the pistons (etc.) to distort. The result will, eventually, be that the engine seizes or the weakest link (e.g. the piston rings, big or little end bearings or one of the con-rods) breaks. The resilience [to stress] of most metals reduces as its temperature increases above a certain temperature.

Car tyres wear very much more quickly, under high acceleration. Cornering too quickly, braking too hard or pulling away from standstill too quickly, all increase the rate of wear on the tyres. This is why Formula 1Grand Prix tyres tend to need replacing after somewhat fewer miles than the tyres on an average saloon driven under normal conditions. It is also why the tyres on *Concorde* tend to wear out more quickly than on most other aircraft – its take-off and landing speeds are higher, so it is necessary to accelerate and decelerate the aircraft more quickly. They also happen to be of a smaller diameter than most other commercial aircraft so it is necessary to get the wheels turning even faster for a given aircraft speed.

In the last section, the notion of measuring the age in stress cycles was introduced. The reason for this is that different users operating over different routes in different conditions will subject the system to different levels of stress. The motorist who spends most of his time driving on motorways at a [relatively] constant speed close to the cruising speed of the vehicle will tend to accumulate far fewer stress cycles per hour/mile than one who drives around town and who is always in hurry. It is not high velocity which causes stress so much as acceleration.

When the Boeing 747 went into service, it was noticed that those being operated by a, mainly internal, Japanese airline were suffering far more failures than those on the long-haul routes (e.g. between UK and USA). The reason for this was that the average stage-length for the particular Japanese operator was less than one hour whereas for the trans-Atlantic routes it was nearer to seven hours. Almost all of the stress cycles on an aircraft are experienced during take-off and landing. Whilst it is cruising at 35,000 ft, say, none of the components are accumulating very much stress. The result of this observation was that Boeing designed a new variant of the '747; the B-747-SP for short-hauls. This had, amongst other changes, strengthened under-carriage system.

A military combat aircraft is seldom flown straight and level, for more than a few seconds. Some pilots will use the rudder and joystick to change direction, others prefer to use variable thrust (by increasing the port engine's thrust and reducing the starboard's to turn right, for example). Both are perfectly satisfactory methods of control but the latter greatly increases the stress count within the two engines. Studies of the stress graphs of the *Red Arrows* (Royal Air Force) aircraft flying a typical aerobatics display show that the No. 9 aircraft (the one at the back) can accumulate 20-30 times as much stress in its engine as the lead aircraft (No. 1). This, of course, means that the No. 9 aircraft is likely to see very much more of the maintenance bay than the No. 1 unless something is done to manage the situation, for example by changing the aircraft around for each mission.

Studies of the stresses on a number of key components within the RB199 Rolls Royce engines on a small sample of *Tornado* aircraft showed that there were a large number of factors which contributed towards the stress accumulation. These included the mission profile, the base from which these missions were flown, the ambient conditions, the aircraft and the engines themselves. It was also suspected that the pilots were major contributors to the variance but this could not be checked due to political reasons.

With the introduction of *full authority digital engine control (FADEC)* systems, the opportunity to fit engine monitoring systems fleet-wide became a practical proposition. These control systems require real-time data from a large number of sources. In most cases, these sources are precisely the same ones required for stress monitoring. It therefore needs little more than some additional software to turn the FADEC into an EMS. Since software is essentially weightless adding such a capability has minimal effect on the performance of the aircraft/propulsion system.

The EMS fitted to one of the latest aircraft collects the values from the various probes at frequencies of up to 16 times per second for the total time that the engines are running. The on-board computer then analyses this and records the cumulative number of cycles. However, this information has to be downloaded at regular intervals or it will be over-written and hence lost. It is therefore necessary to download this data onto a ground-based system where it is recorded against each of the monitored failure modes on each of the safety-critical components.

Since the hard lives of these components are measured in [stress] cycles, it is important (to the safety of the aircraft and crew) that the number of cycles accumulated is accurately recorded. If the data from one, or more, missions has been lost, it is imperative to take this into account. It is also not sufficient to simply fill in the missing missions using the average cyclic-exchange rate (number of cycles per hour) but due cognizance must be taken of the variance. To be safe, it is considered necessary to fill-in any missing values at the upper $\alpha\%$ confidence bound for the values.

The data available is recorded in the form of a number (n) 2-tuples $\{N_i, X_i\}$ where N_i is the number of missions recorded during the ith download and

X_i is the number of cycles counted during the N_i missions. Typically, N_i is likely to be between 3 and 9. The number of counts will vary considerably not only between failure modes but also within the given mode.

Now, since X_i is the sum of the number of stress cycles (counts) from N_i missions, it is not possible to calculate the variance of the number of counts per mission directly. From the Central Limit Theorem (CLT), the variance of a sum is inversely proportional to the sample size or, equivalently, the standard deviation (σ) is inversely proportional to the square root of the sample size. If all the samples were the same size and the number "lost" always consisted of this many missions-worth of counts then there would be no need to estimate the variance in the number of counts per mission. Unfortunately, neither situation prevails – both the number of missions recorded and mission-counts lost vary. The situation is further exacerbated by the fact that each mission is unlikely to be exactly one hour in duration but could vary from 30 minutes to over 4 hours. Given that we know the flying time "lost", then the "fill-in" value should take this into account rather than using the number of missions.

Estimation of a safe "fill-in" rate of cycles per flying hour

Let μ be the [unknown] population mean and σ^2 the [unknown] population variance of the number of cycles per engine-flying hour (C/EFH). All of the downloaded data can be used to estimate the population mean (μ) as

$$E(X) = (\Sigma_{i=1,n} X_i) / (\Sigma_{i=1,n} H_i)$$

Where X_i is the number of cycles accumulated, H_i is the number of EFH flown during the ith download and n is the number of downloads recorded.

The population variance (σ^2) can also be estimated as

$$Var(X) = (\Sigma_{i=1,n} (X_i - H_i E(X))^2 / \max\{1, H_i - 1\}) / (n-1)$$

In Figure (6.8), the estimated standard deviations were calculated using the above formula for three distributions: Exponential ($E_x[5]$), Normal ($N[20, 4^2]$) and Uniform ($U[0,10]$). In each case, 100 downloads consisting of between 1 and 9 sorties were sampled from the respective distribution. This was repeated 10 times for each distribution. The results of these were plotted such that the x-value was the standard deviation for all of the individual sorties and the y-value, the estimated standard deviation using the above formula. The means for each of the three distributions were also plotted in the same way.

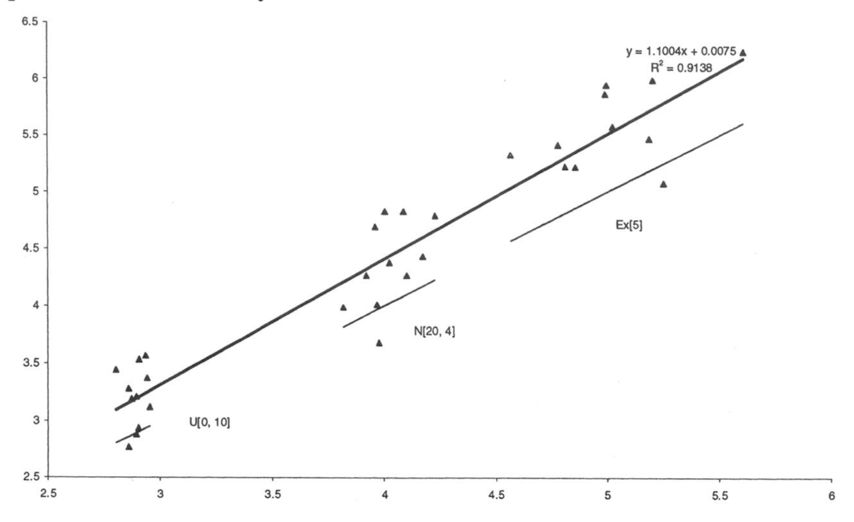

Figure 6-8. Estimated Std Dev v Actual Std Dev for different distributions

The continuous line in Figure 6.8 gives the linear regression line through the 33 points. This shows a slight increase in the biased-ness as the standard deviation increases (slope > 1) although this could be due to the method being more biased when the underlying distribution is Exponential than Normal or Uniform or, perhaps more likely, simply the result of the random numbers used. To determine, with more certainty, which of these is the more likely, it would be necessary to take larger samples over a wider range of distributions.

The broken line in Figure 6.8 is the line "y = x". Any point below this line indicates that the estimated standard deviation was less than the actual. The number of "sorties" making up each of the 100 downloads was sampled from a uniform distribution (U[1,9]). The same number of sorties (553) was used in each trial. The number of "cycles" for each sortie was sampled from the respective distribution. These were then summed over the appropriate number to give the number of cycles per download. In practice, only the number of sorties, the cumulative flying time and the cumulated number of cycles would be known for each download.

In most cases, one would want to use a formula that was unbiased, i.e. the mean of the estimates was very close to the mean of the actual and tending towards this value as the sample size increases. In the case of using a fill-in rate, it is important that the value substituted for any missing data has a high probability of being no less than the actual value which was lost.

The cost of "losing" a few hours-worth of potential life out of a disc whose hard life is likely to be around 2000 hours, or higher, is far less than the cost of the loss of an engine, aircraft, crew or third parties due to an uncontained disc failure (because it was allowed to over-fly its safe life, albeit inadvertently).

6.8 TIME TO FAILURE ANALYSIS

When an engine leaves the factory, its weight, thrust and specific fuel consumption will all have been measured – the last two using a static test-bed. If any of these were outside the acceptable range, the engine would have been rejected and sent back to the assembly shop (or similar facility) to rectify the cause of the non-conformance.

Unfortunately, there is no way of knowing what the "reliability" of the engine, or, indeed, of any of its constituent parts is at this time. Given that the parts have been manufactured to specification (from materials also to specification), then there is a high probability that the so-called "safety-critical" (or to use the jargon, "Group A") parts will not fail before they reach their respective hard lives. This, however, applies to probably less than 1% of the parts within the engine. This is by no means peculiar to aircraft engines; it is equally applicable to the aircraft and all of its components and to any complex system.

The care taken over the manufacturing processes and the level of finish on most components in an aircraft engine is considerably greater than that for most other products and is certainly comparable to that of the electronics industry. Small blemishes on, e.g. turbine blades, can very seriously shorten the time to failure. Particles of swarf or similar debris, can quickly damage bearings and other surfaces particularly taking into account the speeds and temperatures at which these engines operate. A slight imbalance in a shaft or disc can set up vibrations in the engine and gas flow which could make other quite unrelated components to suffer high-cycle fatigue, particularly if the frequency of any of these vibrations happens to coincide with their natural frequency. Similarly, scratches, which can result from poor handling during manufacture, assembly or later during storage or maintenance, may create "stress raisers" which will later become points of weakness from which cracks, etc. may propagate.

There is, to the best of my knowledge, no way of determining the time to failure of any system or component thereof simply from the design. Finite element analysis can help identify the stresses likely to be experienced by any given component under a range of operating conditions. Although this can greatly assist the designers in deciding what materials to use, their thickness and, for example, any radii of curvature, it still only gives a very approximate estimate of the likely time-to-failure.

6.9 ESTIMATING MTTF FROM THE TEST SAMPLE

Testing can help. By running components in a test facility, which [attempts to] simulate the likely stresses incurred in service; until cracks appear or they break, it is possible to build up a sample of times to failure. This sample can then be analyzed to determine which [probability] distribution best describes these times and then the parameters of that distribution which, in some way, provides a best fit to the given times.

Usage of assumptions, for example, that the times will best be described by a 2-parameter Weibull distribution, can reduce the size of the sample needed to achieve a reasonable level of confidence. Making additional assumptions about the shape (i.e. assuming β equals some value b), it is possible to gain considerable confidence about the lower bound of the mean time to failure (or equivalently, the scale, η) from a relatively smaller sample. This method is often referred to in the literature as "Weibayes" or "Weibest" (e.g. Nelson (1982) and Abernethy (1983)).

Supposing N identical components are put on test. At some time T (> 0) the test is terminated or, at least interrupted, before any of the components have failed. At this point, all that is known is that N components have each been "operated" for T hours without failure.

Now suppose that the times to failure for this component can be modeled by a Weibull distribution with shape parameter β and scale parameter η. Further suppose that past experience of similarly designed components indicate that their times to failure can be described by a Weibull distribution with shape β = b but with possibly differing values of the scale parameter, i.e. each of the different designs/systems has the same shape but different scales.

From the maximum likelihood estimation method, Equation 6.7 gives an expression for the estimate of η given β, and T_i where the T_i's are the times to failure for each of the failed components and the times so far achieved at the moment when the experiment was terminated or interrupted.

$$\eta = \left(\frac{\sum_{i=1}^{N} t_i^{\beta}}{r} \right)^{1/\beta} \tag{6.7}$$

Where N is the sample size and r is the number of units within this sample which have failed. In the case when the test has been terminated before any failures have occurred, r = 0 yields an illegal operation in Equation 6.7. To overcome this, it can simply be assumed that one of the N components will fail in the next instance so that now r = 1. This is clearly a pessimistic assumption but it does set a lower limit on the lower bound. In fact, it can be shown that this gives an estimate of η at a confidence level of 63.2% - i.e. there is 63.2% probability that the true mean is greater than that calculated. Indeed, by replacing r with $-\log_e(\alpha)$ (for $0 < \alpha < 1$) the confidence of estimate of the lower bound of η is $100(1-\alpha)\%$. Thus,

$$\Pr\{(\eta > \hat{\eta})\} = \alpha \text{ for}$$

$$\hat{\eta} = \left(\frac{\sum_{i=1}^{N} t_i^{\beta}}{-\log(1-\alpha)} \right)^{1/\beta} \tag{6.8}$$

Using equation (6.8), it is possible to determine the minimum time for which a sample of N components should be tested to be $100\alpha\%$ confident that the scale will be greater than a given value. In engineering, it is more common to use point estimators for B_{10}, B_1 or even $B_{0.1}$ where B_x is the time by which x% can be expected to have failed. But, knowing β and η (or at least the lower bound for η), it is very easy to calculate B_x for the Weibull distribution $W[\beta,\eta]$. The length of time, each of a sample of N components must be run on test, then becomes

$$t_i = \eta \left(\frac{-\log(1-\alpha)}{N} \right)^{1/\beta} \quad \text{for } i = 1, N \tag{6.9}$$

The confidence level of $100\alpha\ \%$ is only achieved if all N components in the sample are still in a SoFu at the end of the test (i.e. when all have achieved an age of at least t). This method is based on the assumption that the value of β is the correct value. It is this last assumption from which this method gets its name, as it is based on Bayes theorem. The validity of this assumption can be tested and, if necessary adjusted by using Bayesian Belief Nets.

A more fundamental problem with relying on test data is that of whether the test conditions are a good representation of the conditions likely to be experienced by the system during future operations. The reason for many systems appearing to be less reliable than the manufacturers and operators had expected has been that the conditions under which the system has had to operate have been somewhat different to those for which it was originally designed. Two examples of this were the Tornado and the main battlefield tank during the Gulf War. In both cases, these systems were designed based on the assumption that a future conflict was most likely to take place on the northern plains of Europe. The most significant difference between this region and that of the Middle East is that the former is largely covered in grass whilst the latter is mainly sand, which can become airborne in large quantities during sand storms / windy conditions. Sand, ingested by the RB199 engines of the Tornado, melted as it passed through the combustion chamber and solidified as glass on the surface of the turbine blades. This stopped the cooling air from passing through these blades with the result that many of them melted causing engine failure. Sand blocked the air filters and got through seals into the engines of the tanks causing them to suffocate and seize.

Both of these systems had undergone significant amounts of testing in their pre-production phase and subsequently in service but, alas, not in the conditions that they were to experience in the Gulf. These are extreme examples but there are many more that could be cited. Another aero-engine that had been in service for many years without too many serious problems underwent a series of modifications. A number of them were tested under various operating conditions satisfactorily but, shortly after entering service, a major fault was discovered. When the aircraft went into a very tight turn, the rotating machinery of the engine rubbed on the stationary parts causing very serious damage. The [static] testing on the engine test beds could not simulate a 9g-turn so did not pick up the fact that the shaft was not strong enough to support these forces without moving the front bearing further forward.

6.10 AGGREGATED CUMULATIVE HAZARD (ACH)

Nevell (2000) introduced a new method for studying times-to-failure. This method, which is expanded on at length in Cole (2001) is particularly good at highlighting, visually, scenarios in which there is a possibility of more than one failure mechanism.

In many cases, particularly within the aero-engine business, small "batches" of components can exhibit significantly different behaviour (in terms of their ttfs) from the remainder of the population. The cause(s) of this untypical behaviour may be due to design changes in the component or, more often, in some other component which may otherwise be unconnected with the one in question. On other occasions, it could be that a previously "dormant" failure mode has come to prominence due to modifications that have reduced the probability of failure to the previously dominant cause.

A classic example of this was when the HP turbine blades in a military engine were redesigned so that they could be grown out of single crystal. This massively increased their MTTF to the point where they rarely caused an engine failure (whereas previously, they were the dominant cause). Unfortunately, the effect of the modification was not as spectacular as was originally hoped because one of the other turbine blades started causing engine failures. The reason this had not been anticipated by the designers was that these blades were very often replaced at the same time as the HP blades (i.e. as secondary rejections) so rarely failed in their own right. It should be noted that in both cases, the ttfs were quite strongly age-related with the Weibull shape parameter $\beta > 3$. It should also be noted that because very few of the more reliable blades had caused engine failures, there was very little data available to determine their ttf distribution.

6.11 PROGNOSTICS

Anything that can tell the operator, or better still, the mechanics when some component will need to be replaced shortly before it becomes imperative has to be beneficial. Ideally, such a mechanism should give sufficient warning to allow the maintenance task to be scheduled into the routine operation of the system. In the case of a commercial aircraft, the ideal would be such that if the warning occurred just after the aircraft had left the [airline's] hub, the task could be safely deferred until it next passed through there which could be 5 to 10 days later, say.

A simple prognostic device fitted to some cars is brake pads constructed with a copper electrode set into the pad at such a position that when it makes contact with the disc, it tells the driver (via a visual warning light on the dashboard) that the pad has only so many millimeters of wear left. Given that the driver recognizes what the warning light means; this allows him to take preventative action before the pads damage the discs.

Other methods rely on monitoring. In 1987, Knezevic (Knezevic 1987a and b) introduced two new concepts in failure management which he called: relevant condition parameters (RCP) and; relevant condition indicators (RCI). He expanded on both of these in his first book (Knezevic 1993).

6.12 RELEVANT CONDITION PARAMETERS RCP

The Relevant Condition Parameter, RCP, describes the condition of an item at every instant of operating time. Usually this parameter is directly related to the shape, geometry, weight, and other characteristics, which describe the condition of the item under consideration. The RCP represents the condition of the item/system which is most likely to be affected by a gradual deterioration failure such as wear, corrosion fatigue crack growth. The general principles of the RCP are discussed by Knezevic (1987). Typical examples of RCP are: thickness of an item, crack length, depth of tyre treads, etc. Saranga (2000) expands at considerable length on the first of these two concepts. She showed that if a component failed due to steady, monotonically increasing level of deterioration then given this could be monitored, it could be used to improve the predictability of when the component (and hence, possibly, the system) would fail.

A particularly common application of this technique is when motorists inspect their tyres. The depth of tread is a monotonically decreasing function of usage. Having an estimate of the rate of wear (and the amount of tread remaining) makes it possible to adjust the times between inspections. The same inspection can also identify whether the tracking and/or alignment is causing excessive wear on the inside or outside of the tyres. In this case, tyres tend to wear very much more quickly and can cause vibration, usually at fairly high speeds, if left unattended for too long.

Another application of this technique is with pipes, particularly those carrying corrosive fluids or exposed to corrosive or erosive substances. Over a period of years, the thickness of the pipes' walls will decrease to a point, ultimately, when the pipe is unable to contain the fluid passing through it. Determining whether a pipe is in danger of imminent rupture can be an expensive process, possibly involving digging up long stretches of the highway or sending divers or submersibles to the seabed. By knowing the original thickness and measuring it again after a given period, it is possible to estimate the rate of material (thickness) loss and hence predict when the pipe will need to be replaced. By analyzing the debris scavenged from the fluid as it passes through filters, it may also be possible to measure the amount of material being lost from the inner surface of the pipe.

This latter method is being used very effectively in aero-engines. The lubricant/coolant (oil) passing over the bearings and other friction-bearing parts naturally collects the microscopic particles of metal that detach from these components during usage. Collecting these particles and subjecting them to electron microscope scanning and spectroscopic analysis allows the inspector to identify the sources and measure the amount of erosion occurring. Failure is often preceded by an increase in either the quantity or the size of the particles being scavenged in this way, although a monitor of the steady accumulation may help to predict the onset of failure sufficiently far in advance to allow preventative maintenance to be planned at the operator's convenience.

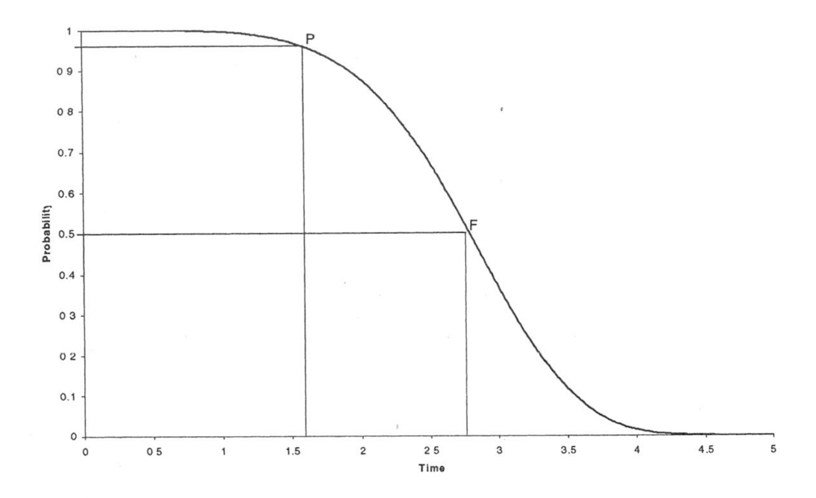

Figure 6-9. P-F Curve for an RCP

Figure 6.9 shows the traditional picture used by proponents of reliability centred maintenance (RCM) which shows how the value of a performance parameter declines with age. The point P is the earliest time at which deterioration can be detected. At some time later, at point F, the component/system has deteriorated to the point where maintenance is necessary. Using Knezevic's definitions of relevant condition parameters and indicators, Figure 6.9 is an example of an RCP. As such, it is unlikely to be the graph of a performance metric as it is too smooth, although it could represent the underlying shape upon which random variation or noise would be superimposed.

If we take the depth of tread on a car tyre as our "measure of performance" then it could be argued that the point at which "deterioration" starts would be at time $t = 0$ and, this would also correspond with the point at which this deterioration is first detectable. Failure, at least in the UK, is defined as the time when the depth of tread at some area on the tyre is less than 1.6mm which might represent a "y" value of around 0.16 (assuming the original depth was 10mm) or a time of somewhere approaching 3.5.

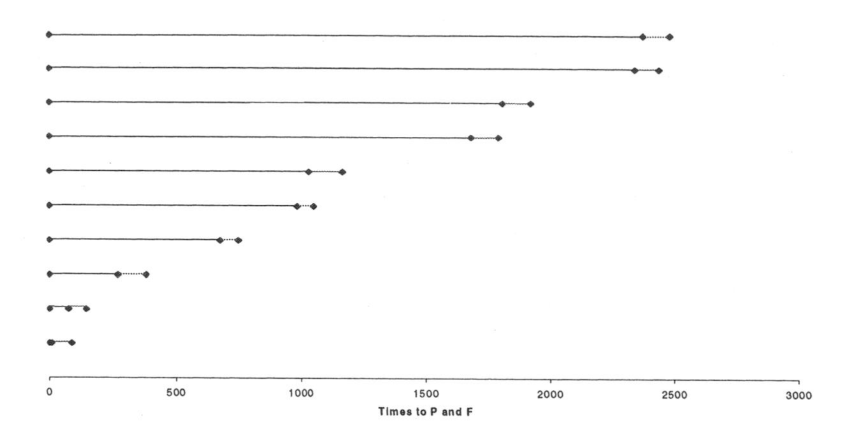

Figure 6-10. Time to P and F

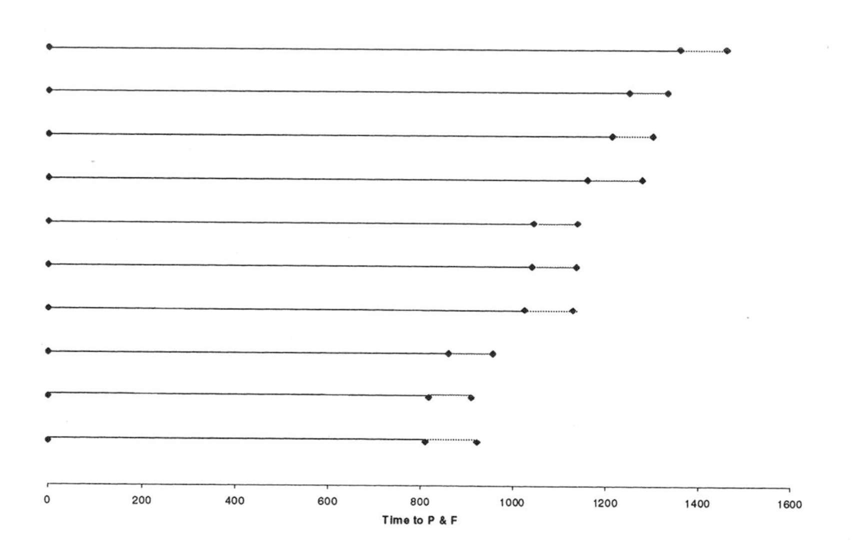

Figure 6-11. Time to P and F

In the case of water pipes, the thickness of pipe wall might start at 5mm when new. If it is known that a pipe whose walls were at least 2.5 mm thick can withstand the likely maximum stresses of ground movements (from traffic and weather, but maybe not from earth quakes) then this could be chosen as point P. In this case, it could be argued that there is no need to inspect the pipes at frequent intervals until the wall thickness has halved (assuming there is a reasonable estimate of the rate of loss of wall thickness due to erosion and corrosion). If the pipe is liable to fracture at any time if the thickness drops below 1.5mm, say, then this would represent point F.

Given a "performance measure" is the length of a crack in a wing spar, suspension mounting or turbine disc then it is to be hoped that P will be a long time after the system went into service. In the case of a turbine disc in an engine on a commercial airliner, the time to P could be of the order of 100,000 engine flying hours, for the wing spar, possibly in excess of 250,000 flying hours and the suspension mounting maybe 150,000 miles. However, once the crack starts, the time to failure is likely to be very much shorter. Figures 6.10 and 6.11 illustrate two sets of ten such systems. In Figure 6.10 the times to start crack (P) were sampled from an exponential distribution with λ = 1000 and the times from P to F from a Weibull distribution W[5, 100]. The first started to crack after just 8 hours and failed 80 hours later whilst the tenth started to crack at 2375 hours and lasted 108 hours before failing. By contrast, in Figure 6.11, the times to start crack (P) were sampled from a Weibull distribution W[5, 1089] whilst the crack propagation times (P-F) were sampled from the same distribution as before. In this case the shortest time to P was 810 hours and the longest, 1360 hours. [Note: the reason for the rather strange value of η (1089) was so that the mean time to failure was the same for both cases, i.e. 1000 hours.]

Given these two distributions for Case 1 – an MTBF = 1000 hours and a crack propagation time ~W[5, 100] hours – the first question is when to start routine inspections and secondly how often to repeat them to have a given probability of detecting a crack before it causes a failure. Suppose a 10% probability of a failure is acceptable, then inspection should start at:

$$0.1 = 1 - e^{-\frac{t}{1000}}$$
$$\text{giving} \tag{6.10}$$
$$t = -1000\ln(0.9) = 105$$

After this time, continuous monitoring with a 100% probability of detection would be necessary to achieve a probability of failure of no greater than 10%. The alternative is to consider the probability that the crack starts in a given interval ($T_{i-1} < P < T_i$) and fails before the end of the same interval at time $t < T_i$, i.e. $F\text{-}P < I = T_i - T_{i-1}$. Assuming that point P can occur at any time during the inspection interval with equal probability then:

$$\Pr\{F - P < I\} = \frac{1}{I}\int_0^I F_2(I - x)dx$$

$$= \frac{1}{I}\int_0^I \left(1 - e^{-\left(\frac{I-x}{\eta}\right)^{\beta}} \right) dx \qquad (6.11)$$

From the graph in Figure 6.12, it can be seen that to be 90% confident of detecting a crack before it has chance to cause a failure, the inspection interval will need to be around 95 hours starting from when the component is new. To increase the probability of detection to 95% the inspection interval would have to reduce to 80 hours.

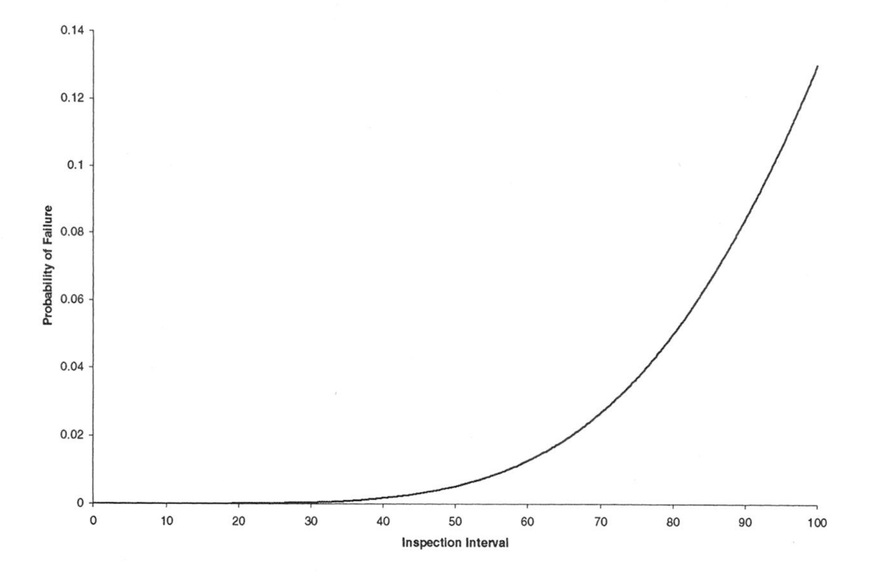

Figure 6-12. Graph of the probability of failure against the inspection interval

Now, for Case 2 the propagation time has the same distribution but a rather different time to start crack. The B_5 [B_{10}] for W[5, 1089] is 601 [694] hours. This means the start of the inspection regime can be delayed until the component's age is 600 [700] hours. After this time, inspection would need to be performed every 80 [95] hours to be 95% [90%] confident of finding the crack before the component fails.

6.13 RELEVANT CONDITION INDICATORS RCI

Whereas RCP tend to operate best at the part level, RCI more commonly apply at the system or sub-system level. Unlike the RCP, RCI may not be monotonically decreasing (or increasing) rather they exhibit trends which taken over time may show an underlying downwards (or upwards) movement.

A typical RCI is fuel consumption. When an engine is performing at its peak, the fuel consumption will be at its lowest. However, as sparkplug gaps increase, valves and piston rings wear and HT leads breakdown, the fuel consumption (measured in miles per gallon – mpg or litres per kilometer – lpk) will tend to decrease (i.e. the amount of fuel used for a given distance will tend to increase). The actual value will show considerable variation on a day-to-day basis as weather, driving conditions and routes change.

The difficulty with RCI as monitors of condition is that it is often very difficult to extract the underlying trend from the day-to-day noise. Methods such as moving average, exponential smoothing and Kalman filtering are useful aids but obviously depend on the "signal-to-noise" ratios. The other problem with this method is deciding the criteria for defining when the system has deteriorated below an acceptable level.

Most cars, for example, have recommended regular service intervals (of miles covered) although some more expensive models now have on-board monitoring systems which advise the driver of the "best" time. The [old] interval is probably based on a combination of tradition and some expectation of driver behaviour. The fact that most cars have the same interval, suggests that it has little to do with the type of car, the reliability of its components or the performance of the engine. An obvious exception is a Formula 1 racing car which rarely achieves anything like the normal service interval of 6000 miles – more likely limit being nearer to 300 miles for a complete overhaul (not just oil change), according to Corbyn (2000).

In Figure 6.13 the smooth line represents a typical RCP, that is, a monotonically decreasing function of time (x-axis) whilst the jagged line shows what an RCI plot of condition against time may look like. Suppose a criterion for performing (preventative) maintenance was when the measure of "condition" fell below 0.8, then this would be at t = 2.23 based on the RCP but could be as early as t = 0.22 if using the RCI (as shown). Using exponential smoothing with a damping factor of 0.8 (i.e. $X_{i+1} = 0.8X_i + 0.2RCI_{i+1}$ where RCI_i is the value of the ith RCI value) gives the horizontal lines which first crosses the value y = 0.8 at t = 2.19. Obviously, different values of the damping factor would produce different times.

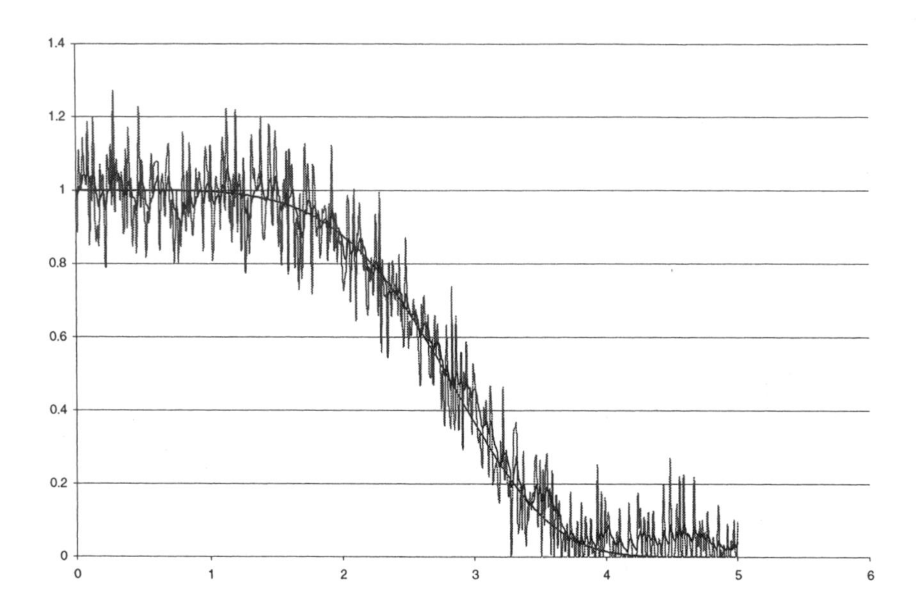

Figure 6-13. Comparison of an RCP with an RCI

6.14 CASE STUDY: AIRCRAFT ENGINE COMPONENTS

The data for this case is taken from a medium sized commercial airline flying to over 25 destinations within Europe. It flies over 1000 flights a week, with average flight time of around 1 hour and the maximum flight to any of their destinations is 2 hours. On average, it carries over 3.2 million passengers per annum. Our analysis is restricted to a particular type of aircraft (we call it aircraft Alpha), and the relevant information on its characteristics and usage are as follows:

Aircraft	Alpha
Fleet Size	16
Capacity	70 – 110 Seats
Number of flights	1000 per week
Number of destinations	25

Speed	400 Knots
Altitude	30,000 feet
Engine	Gas Turbine
Number of Engines	4
Flight Time	45 minutes to 2 hours

6.14.1 The data

Our first task is to identify the components that cause the majority of the unscheduled maintenance. The data used in the case study are collected over a period of three years and is a pooled data set over 16 aircraft. Each datum corresponds to the removal of a component due to some reported fault. The ages at removal of all the components are measured in terms of flying hours. The data also consists of age in terms of calendar hours and number of cycles (landings). The data does not include any planned removals or censored data. The components are subjected to predictive maintenance. The data consist of following entries:

1. Part Number
2. Description
3. Aircraft
4. Position (Engine)
5. Date ON/OFF
6. Number of flying hours before the component removal
7. Number of flights before component removal
8. Original Equipment Manufacturer (OEM)
9. Reason for removal (Initial Fault)
10. Repair agency strip report finding
11. FC/FF/NFF (Fault confirmed by the repair agency / Fault not confirmed but found another fault / No fault found).

The important entries for this case study are entry numbers 6, 7, 9, 10 and 11. Several thousands removal reports were analyzed to identify the components that caused the majority of the unscheduled maintenance; the top 10 components that caused the majority of the unscheduled maintenance are listed in Table 6.2 along with their time to removal distribution. The time to removal distribution is found using the software PROBCHAR™ (Knezevic, 1995).

Table 6-2. Top 10 components responsible for unscheduled maintenance and their Time to removal distribution

Part Name	Distribution	Parameter Estimates
TMS Actuator	Weibull	$\hat{\eta} = 778,\ \hat{\beta} = 0.7$
Attitude Direction Indicator	Weibull	$\hat{\eta} = 365,\ \hat{\beta} = 0.65$
Fuel flow indicator	Weibull	$\hat{\eta} = 1700,\ \hat{\beta} = 0.85$
Vertical Ref Gyro	Weibull	$\hat{\eta} = 890,\ \hat{\beta} = 1.12$
Altitude Selector	Weibull	$\hat{\eta} = 375,\ \hat{\beta} = 0.78$
Central Audio Unit	Exponential	$1/\hat{\lambda} = 1020$
Navigation Controller	Exponential	$1/\hat{\lambda} = 1420$
Pressure Transmitter	Weibull	$\hat{\eta} = 219,\ \hat{\beta} = 0.98$
Flight Data Recorder	Exponential	$1/\hat{\lambda} = 1280$
Auto Pilot Computer	Exponential	$1/\hat{\lambda} = 1050$

For TMS Actuator, Attitude direction indicator, Fuel flow indicator, Vertical Ref Gyro, Altitude Selector and Pressure transmitter, the cumulative distribution function of time to removal random variable selected is the Weibull distribution, given by:

$$F(t) = 1 - \exp(-(\frac{t}{\eta})^{\beta}) \qquad \alpha, \beta > 0 \tag{6.12}$$

For Central audio unit, Navigation controller, Flight data recorder and auto pilot computer, the cumulative distribution function of time to removal random variable selected is the exponential distribution given by:

$$F(t) = 1 - \exp(-\lambda t) \qquad t > 0, \lambda > 0 \tag{6.13}$$

Here η (in flying hours) and β denote the scale and shape parameters of the Weibull distribution, respectively. The notation λ is used to represent the failure rate per flying hour (thus $1/\lambda$ would give us the mean time between removals). The surprising factor here is that many of these components have shape parameter value less than 1. This corresponds to a decreasing failure rate and normally indicates a quality problem with possibly insufficient or inadequate pass-off or burn-in testing (i.e. running the component on a test rig prior to delivery to check for manufacturing or assembly faults). It should be also noted that the random variable under consideration is time-to-removal and not time-to-failure. Although the components were removed based on some fault being reported, this did not actually record the cause. The data also did not indicate whether any modifications had been made so it was not possible to determine whether the times were from a homogenous set.

6.14.2 Analysis of failures generating unscheduled maintenance

Unscheduled maintenance can easily wipe out the profit of an airline. Delay and cancellation costs have become so high that no airline can afford to take any chances. The delay cost in the case of Boeing 747 is estimated to be in the order of $1000 per minute. It is also estimated that around 28% of the delays are caused by technical problems that lead to unscheduled maintenance (Knotts, 1996).

The expected number of unscheduled maintenance actions can be derived as follows:

Exponential time-to-removal distribution:

For the exponential time-to-removal distribution, the expected number of unscheduled removals for a period of operation t, $UR_e(t)$, is given by the well known expression (in this case the removals form a Poisson process with mean λt):

$$UR_e(t) = \lambda t \tag{6.14}$$

Weibull time-to-removal distribution:

When the time-to-removal is Weibull, we use the renewal function to find the expected number of removals for an operating period of t hours. The expected number of unscheduled maintenance, $UR_w(t)$, is given by (assuming that after removal, the components are replaced with components that are '*as-good-as-new'*, which is seldom true):

$$UR_W(t) = \sum_{n=1}^{\infty} F^n(t) \tag{6.15}$$

Where $F^n(t)$ is the n-fold convolution of the time-to-removal distribution $F(t)$ given in equation (6.12). Equation (6.15) is difficult to solve analytically. For this purpose we use the iterative method proposed by Xie (1989), by exploiting another renewal function given by:

$$UR_w(t) = F(t) + \int_0^t UR_w(t-x)dF(x) \tag{6.16}$$

Now by setting $t = nh$, that is, the period t is divided into n equal intervals of size h, equation (6.16) can be written as:

$$UR_w(nh) = \frac{1}{1-(h/2)f(0)}[f(nh) + \frac{h}{2}UR_w(0)f(nh) + h\sum_{k=1}^{n-1} UR_w[(n-k)h]f(kh)] \tag{6.17}$$

Here $f(t)$ is the probability density function of the Weibull distribution, and is given by:

$$f(t) = \frac{\beta}{\eta}\left(\frac{t}{\eta}\right)^{\beta-1} \exp\left[-\left(\frac{t}{\eta}\right)^{\beta}\right]$$

Using equations (6.14) and (6.17), we can find the number of unscheduled maintenance tasks per year. The estimated number of unscheduled maintenance tasks per year (approximately 3000 flying hours per aircraft) for the top 10 components are listed in Table 6.3. The observed numbers of unscheduled maintenance actions over a three-year period were close to the estimated numbers.

Table 6-3. Unscheduled maintenance tasks per year

Part Name	Number of unscheduled maintenance tasks
TMS Actuator	50
Attitude Direction Indicator	97
Fuel flow indicator	26
Vertical Ref Gyro	56
Altitude Selector	110
Central Audio Unit	47
Navigation Controller	34
Pressure Transmitter	219
Flight Data Recorder	38
Auto Pilot Computer	46

The total number of unscheduled maintenance per year due to just 10 components alone (of course, for a fleet of 16 aircraft) is 723. This also represents 723 potential delays/cancellations due to 10 components for a fleet of 16 aircraft. This is very high considering the fact that a modern jet aircraft has up to 4 million components (of which around half are fasteners) and over 135,000 unique components. We also noticed a large percentage of no fault found in some of the components, which is the main discussion point in the next section.

6.14.3 No fault found (NFF)

No fault found, NFF, is basically a reported fault for which subsequently no cause can be found. Isolating the true cause of failure of a complex system naturally demands a greater level of analytical skill, particularly where there is a fault ambiguity present. If the technical skills cannot resolve a failure to a single unit then the probability of making errors of judgment will increase, depending on the level of ambiguity (Chorley,

1998). This problem is not unique to the Alpha airline. Data presented by Knotts (1994) quotes a Boeing figure of 40% for incorrect part removals from airframes, and British Airways estimate that NFF cost them on the order of £20million (GBP) ($28million) per annum.

The percentage of NFF on the top 10 components is listed in Table 6.4, which was obtained from the database of the airline under investigation. Analysis of the NFF data also yielded the following:

1. Around 19% of the unscheduled removals were NFF.
2. Four aircraft accounted for 49% of the NFF.
3. Around 65% of the auto pilot computers that were NFF came from a particular aircraft. All but one of these removals were carried out at a particular maintenance base.
4. Around 45% of NFF removals came from two particular bases.

In a few cases, we also noted that many of the NFF occurred within a few days of each other, for example, there were four consecutive Actuator removals that occurred within two weeks. Similarly, four auto pilot computers were removed from a particular aircraft within 45 days. Understanding the reasons for NFF is more complicated, as it demands a complete scrutiny of the maintenance process. Nevertheless, it is worth trying to understand the causes of NFF, as it can cost around $1000 to $3000 for a note reading "NFF" and a wait of as long as 60 days, depending on the contractor or procedure (Morgan, 1999). Chorley (1998) lists the following external factors that may cause the level of no fault found:

1. Quality and depth of training.
2. Quality of technical data.
3. Test equipment suitability, accuracy and calibration.
4. Design of BIT, its resolution of ambiguities and clarity.
5. Intermittent faults.
6. Human stress and fatigue induced error.
7. Willful intent.
8. The increasing impact of software.

In due course, technology may provide truly intelligent products that not only detect a fault but also are capable of analyzing the actual cause and performing self-repair. In fact, this is already available in many computers. Recently *Airbus* devised a method in which an item that is found NFF by the maintenance agency is labeled as a *rough* item, and is closely monitored to unearth the cause.

Table 6-3. Percentage NFF on Components

Part Name	% NFF removals
TMS Actuator	26%
Attitude Direction Indicator	21%
Fuel flow indicator	31%
Vertical Ref Gyro	26%
Altitude Selector	21%
Central Audio Unit	28%
Navigation Controller	35%
Pressure Transmitter	20%
Flight Data Recorder	18%
Auto Pilot Computer	6%

Chapter 7

RELIABILITY AND SIX SIGMA ESTIMATION

The fewer the facts, the stronger the opinion.
Arnold H Glasow

7.1 INTRODUCTION

To predict various reliability and Six Sigma measures of a product, it is essential that we have sufficient information on the time to failure characteristics of that item. In most cases these characteristics are expressed using theoretical probability distributions. The selection of the appropriate probability distribution function to describe the empirical data plays a critical role in reliability analysis. Once the distribution is identified, then one can extract information about other reliability characteristics such as mean time between failures, hazard function and failure rate etc.

To begin with we look at ways of fitting probability distributions to in-service data, that is, the data relating to the age of the components at the time they failed while they were in operation. In the literature there are three popular tools available to find the best distribution that describes the in-service data: 1. Probability papers, 2. linear regression, and 3. Maximum likelihood estimator. In this book, we discuss linear regression and maximum likelihood estimator techniques. Very often we do not have a complete set of failure data; a section on censored data describes how such data can be used to predict reliability characteristics.

Components may fail due to a number of failure modes. Each of these modes may be more or less related to the age. One would not expect corrosion to be the cause of failure during the early stages of the component's life, unless it was subjected to exceptionally corrosive chemicals. On the other hand, if the components have been badly made then one might expect to see them fail soon after the unit has been assembled.

In this chapter, we first look at the empirical approaches for finding estimates for failure function, reliability function and MTTF. Rest of the chapter describes some of the well-known methods for selection of the most relevant theoretical distribution functions for the random variables under consideration.

7.2 RELIABILITY ESTIMATION AND FAILURE DATA

A common problem in reliability engineering is the availability of failure data. In many cases getting sufficient data for extracting reliable information is the most difficult task. This may be due to the fact that there is no good procedure employed by the operator (or supplier) to collect the data or the item may be highly reliable and the failure is very rare. However, even without any data, one should be able to predict the time-to-failure distribution if not the parameters. For example, if the failure mechanism is corrosion, then it cannot be an exponential distribution. Similarly if the failure cause is 'foreign object damage' then the only distribution that can be used is exponential. The main problem with insufficient failure data is getting an accurate estimate for the shape parameter. If a component is being used in a number of different systems, it may be reasonable to assume that the failure mechanism in each of these instances will be similar. Even though the way the different systems operate may be different, it is still likely that the shape of the failure distribution will be the same and only the scale will be different. The reliability data can be obtained from the following sources:

1. Field data and the in-service data from the operator using standard data capturing techniques. *Failure reporting* forms are available for the purpose of capturing desired information regarding the reliability of the item under consideration. Unfortunately, all these forms are flawed, as they record only MTBF. Just the value of MTBF alone may not be enough for many analyses concerning reliability.
2. From *life testing* that involves testing a representative sample of the item under controlled conditions in a laboratory to record the required data. Sometimes, this might involve '*accelerated life testing*' (ALT) and '*highly accelerated life testing*' (HALT) depending on the information required.

As mentioned earlier, in some cases it is not possible to get a complete failure data from a sample. This is because some of the items may not fail during the life testing (also in the in-service data). These types of data are called '*censored data*'. If the life testing experiment is stopped before all the items have failed, in which case only the lower bound is known for the items that have not failed. Such type of data is known as '*right censored data*'. In few cases only the upper bound of the failure time may be known, such type of data is called '*left censored data*'.

7.3 ESTIMATION OF PARAMETERS – EMPIRICAL APPROACH

The objective of empirical approach is to estimate failure function, reliability function, hazard function and MTTF from the failure times. In the following sections we discuss methods for estimating various performance measures used in reliability estimation from different types of data.

7.3.1 Estimation of parameters – Complete Ungrouped Data

Complete ungrouped data refers to a raw data (failure times) without any censored data. That is, the failure times of the whole sample under consideration are available. For example, let t_1, t_2, ..., t_n, represents n ordered failure times such that $t_i \leq t_{i+1}$. Then the possible estimate for failure function (cumulative failure distribution at time t_i) is given by:

$$\hat{F}(t_i) = \frac{i}{n} \tag{7.1}$$

Where i is the number of units failed by time t, out of the total n units in the sample. This will make $F(t_n) = n/n = 1$. That is, there is a zero probability for any item to survive beyond time t_n. This is very unlikely, as the times are drawn from a sample and it is extremely unlikely that any sample would include the longest survival time. Thus the equation (7.1) underestimates the component survival function. A number of mathematicians have tried to find a suitable alternative method of estimating the cumulative failure probability. These range from using n+1 in the denominator to subtracting 0.5 in the numerator and adding 0.5 in the denominator. The one that gives the best approximation is based on *median rank*. Bernard's approximation to the median rank approach for cumulative failure probability is given by

$$\hat{F}(t_i) = \frac{i - 0.3}{n + 0.4} \tag{7.2}$$

Throughout this chapter we use the above approximation to estimate the cumulative distribution function or failure function. From equation (7.2), the estimate for reliability function can be obtained as

$$\hat{R}(t_i) = 1 - \hat{F}(t_i) = 1 - \frac{i - 0.3}{n + 0.4} = \frac{n - i + 0.7}{n + 0.4} \tag{7.3}$$

The estimate for the failure density function $f(t)$ can be obtained using

$$\hat{f}(t) = \frac{\hat{F}(t_i) - \hat{F}(t_{i+1})}{t_i - t_{i+1}}, \quad t_i \le t \le t_{i+1} \tag{7.4}$$

Estimate for the hazard function can be obtained by using the relation between the reliability function $R(t)$ and the failure density function $f(t)$. Therefore,

$$\hat{h}(t) = \hat{f}(t) \Big/ \hat{R}(t) \quad \text{for} \quad t_i < t < t_{i+1} \tag{7.5}$$

An estimate for the mean time to failure can be directly obtained from the sample mean. That is,

$$\hat{MTTF} = \sum_{i=1}^{n} \frac{t_i}{n} \tag{7.6}$$

Estimate for the variance of the failure distribution can be obtained from the sample variance, that is

$$s^2 = \sum_{i=1}^{n} \frac{(t_i - \hat{MTTF})^2}{n - 1} \tag{7.7}$$

7.3.2 Confidence Interval

It is always of the interest to know the range in which the measures such as MTTF might lie with certain confidence. The resulting interval is called a *confidence interval* and the probability that the interval contains the estimated parameter is called its *confidence level* or *confidence coefficient*. For example, if a confidence interval has a confidence coefficient equal to 0.95, we call it a 95% confidence interval. To derive a (1-α) 100% confidence interval for a *large sample* we use the following expression:

$$\hat{MTTF} \pm z_{\alpha/2}(\frac{\sigma}{\sqrt{n}}) \tag{7.8}$$

Where $z_{\alpha/2}$ is the z value (standard normal statistic) that locates an area of $\alpha/2$ to its right and can be found from the normal table. σ is the standard deviation of the population from which the sample was selected and n is the sample size. The above formula is valid whenever the sample size *n* is greater than or equal to 30. The 90%, 95% and 99% confidence interval for MTTF with sample size $n \geq 30$ are given below:

$$\text{90\% confidence } \hat{MTTF} \pm 1.645 \times \left(\frac{\sigma}{\sqrt{n}}\right) \tag{7.9}$$

$$\text{95\% confidence } \hat{MTTF} \pm 1.96 \times \left(\frac{\sigma}{\sqrt{n}}\right) \tag{7.10}$$

$$\text{99\% confidence } \hat{MTTF} \pm 2.58 \times \left(\frac{\sigma}{\sqrt{n}}\right) \tag{7.11}$$

When the sample size is small (that is when n is less than 30), the confidence interval is based on t distribution. We use the following expression to calculate (1-α)100% confidence interval.

$$\hat{MTTF} \pm t_{\alpha/2}\left(\frac{s}{\sqrt{n}}\right) \tag{7.12}$$

Where $t_{\alpha/2}$ is based on (n-1) degrees of freedom and can be obtained from *t* distribution table (see appendix).

EXAMPLE 7.1

Time to failure data for 20 car gearboxes of the model *M2000* is listed in Table 7.1. Find:

1. Estimate of failure function and reliability function.
2. Plot failure function and the reliability function.
3. Estimate of MTTF and 95% confidence interval.

Table 7-1. Failure times of gearboxes (measured in miles)

1022	1617	2513	3265	8445
9007	10505	11490	13086	14162
14363	15456	16736	16936	18012
19030	19365	19596	19822	20079

The failure function and reliability function can be estimated using equations 7.2 and 7.3. Table 7.2 shows the estimated values of failure function and reliability function. The failure function and the reliability function graph are shown in Figures 7.1 and 7.2 respectively. The estimate for mean time to failure is given by:

$$\hat{MTTF} = \sum_{i=1}^{20} \frac{t_i}{20} = 12725.5 \text{ miles.}$$

Estimate for the standard deviation is given by

$$\hat{s} = \sqrt{\sum_{i=1}^{n} \frac{(t_i - \hat{MTTF})^2}{n-1}} = 6498.3 \text{ miles}$$

As the sample data is less than 30, we use equation (7.12) to find the 95% confidence level. From t-table the value of $t_{0.025}$ for (n-1) = 19 is given by 2.093. The 95% confidence level for MTTF is given by:

$$\hat{MTTF} \pm t_{\alpha/2}\left(\frac{s}{\sqrt{n}}\right) = 12725.5 \pm 2.093(6498.3/\sqrt{19})$$

That is, the 95% confidence interval for MTTF is (9605, 15846).

Table 7-2. Estimate of failure and reliability function

Failure data	$\hat{F}(t_i)$	$\hat{R}(t_i)$
1022	0.0343	0.9657
1617	0.0833	0.9167
2513	0.1324	0.8676
3265	0.1814	0.8186
8445	0.2304	0.7696
9007	0.2794	0.7206
10505	0.3284	0.6716
11490	0.3774	0.6225
13086	0.4264	0.5736
14162	0.4754	0.5246
14363	0.5245	0.4755
15456	0.5735	0.4265
16736	0.6225	0.3775
16936	0.6716	0.3284
18012	0.7206	0.2794
19030	0.7696	0.2304
19365	0.8186	0.1814
19596	0.8676	0.1324
19822	0.9167	0.0833
20079	0.9657	0.0343

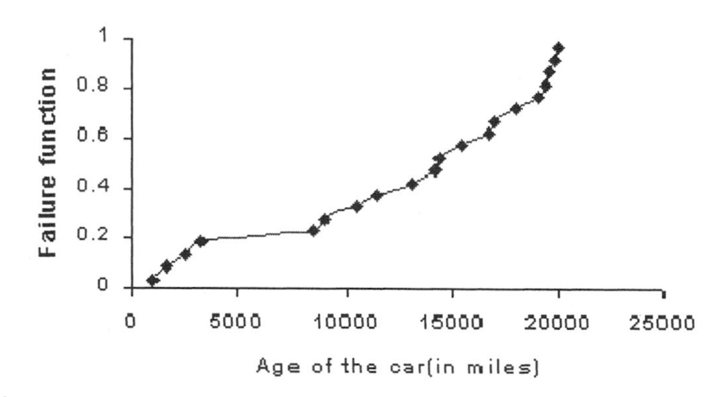

Figure 7-1. Estimate of failure function for the data shown in Table 7.1

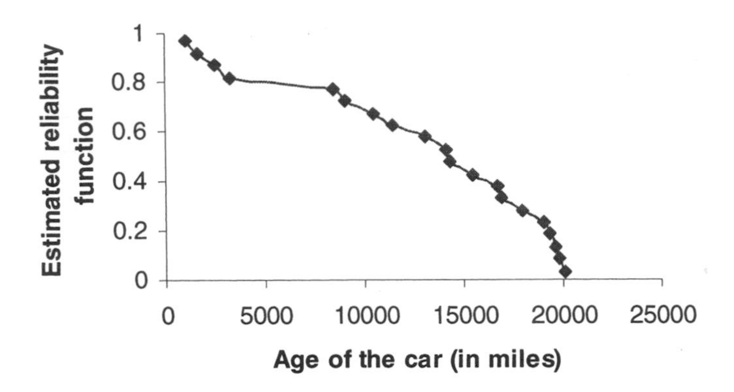

Figure 7-2. Estimate of reliability function for the data shown in Table 7.1

7.3.3 Analysis of grouped data

Often failure data is placed into time intervals when the sample size is large. The failure data are classified into several intervals. The number of intervals, *NI*, depends on the total number of data *n*. Following equation can be used as guidance for determining the suitable number of intervals:

$$\lfloor NI \rfloor = 1 + 3.3 \times \log_{10}(n) \tag{7.13}$$

The symbol $\lfloor \; \rfloor$ denotes that the value is rounded down to the nearest integer.

The length of each interval, *LI*, is calculated using:

$$LI = \frac{(x_{\max} - x_{\min})}{\lfloor NI \rfloor} \tag{7.14}$$

Where x_{max} is the maximum recorded failure time and x_{min} is the minimum recorded failure time. The lower and upper bounds of each interval are calculated as follows:

$$X_{\min,i} = x_{\min} + (i-1) \times LI$$
$$X_{\max,i} = x_{\min} + i \times LI$$

$X_{min,i}$ is the lower bound of the ith interval and $X_{max,i}$ is the upper bound value of the ith interval. Let n_i, $i = 1,\dots,n$ be the number of items that fail in the interval i. Then the estimate for cumulative failure distribution is given by

$$\hat{F}(X_{\max,i}) = \frac{\sum_{k=1}^{i} n_k - 0.3}{n + 0.4} \tag{7.15}$$

Estimate for the reliability function R(t) is given by:

$$\hat{R}(X_{\max,i}) = 1 - \hat{F}(X_{\max,i}) = \frac{\sum_{k=i+1}^{n} n_i + 0.7}{n + 0.4} \tag{7.16}$$

Estimate for the failure density is given by:

For $X_{max,i+1} < t < X_{max,i}$

$$\hat{f}(t) = \frac{\hat{F}(X_{\max,i+1}) - F(X_{\max,i})}{X_{\max,i+1} - X_{\max,i}} = \frac{n_{i+1}}{(n+0.4) \times (X_{\max,i+1} - X_{\max,i})}$$

The MTTF is estimated using the expression:

$$\hat{MTTF} = \sum_{i=1}^{NI} \frac{X_{med,i} \times n_i}{n} \tag{7.17}$$

Where $X_{med,i}$ is the midpoint in the ith interval and n_k is the number of observed failures in that interval. Estimate for sample variance is given by

$$s^2 = \sum_{i=1}^{NI} (X_{med,i} - \hat{MTTF})^2 \times \frac{n_i}{n} \tag{7.18}$$

EXAMPLE 7.2

Results of 55 observed values of failure times of batteries used in locomotives (in months) are given in Table 7.3. Calculate the estimate of Mean Time to Failure (MTTF).

Table 7-3. Failure times of batteries measured in months

3	56	9	24	56	66	67	87	89	99	4
26	76	79	89	45	45	78	88	89	90	92
99	2	3	37	39	39	77	93	21	24	29
32	44	46	5	46	46	99	47	77	79	89
31	78	34	66	86	86	75	33	55	22	44

First we need to find the number of groups using equation (7.13). The number of intervals is given by:

$$\lfloor NI \rfloor = 1 + 3.3 \times \log_{10}(55) = \lfloor 6.74 \rfloor = 6$$

The length (range) if each interval (group) is given by:

$$LI = \frac{x_{\max} - x_{\min}}{\lfloor NI \rfloor} = \frac{99-2}{6} = 16.17$$

Table 7.4 shows the various calculations associated in computing the mean time to failure. The estimate of MTTF is given by:

$$\hat{MTTF} = \sum_{i=1}^{NI} \frac{X_{med,i} \times n_i}{n} = \sum_{i=1}^{6} \frac{X_{med,i} \times n_i}{55} = 55.06 \text{ Months}$$

Table 7-4. Analysis of grouped data

i	LI ($x_{min,I}$ - $x_{max,i}$)	n_i	$x_{med,i}$	$X_{med,i} \times n_i$
1	2 - 18.17	6	10.08	60.51
2	18.17 – 34.34	10	26.25	262.55
3	34.34 - 50.51	11	42.42	466.67
4	50.51 - 66.68	5	58.59	292.97
5	66.68 - 82.85	9	74.76	672.88
6	82.85 - 99	14	90.92	1272.95

7.3.4 Analysis of censored data

In many cases, the complete data may not be available due to the reasons such as all the items may not have failed or the manufacturer may wish to get interim estimates of the reliability etc. The mechanism for censoring may be based on a fixed age, on a fixed number of failures or at some arbitrary point in time. In practice, assuming the times at the time of failure or at the time of suspension (censor) are known, the reason for terminating the test is not important. We will assume that the times of failure and suspension are known precisely. In this section we derive estimates for failure function and reliability function where the data is censored. We use t_i, to represent time in case of complete data and t_i* to denote a censored time.

The only difference between the estimation of parameters in complete data and the censored data is the calculation of median ranks. Now we will need to adjust the ranks in order to take account of the components that have not failed. The rank adjustment is done in the following two steps:

1. Sort all the times (failures and suspensions) in ascending order and allocate a sequence number i starting with 1 for the first (lowest) time to failure (suspension) and ending with n (the sample size for the highest recorded time). Now we discard the suspended times as it is only the (adjusted) rank of the failures we are concerned about.
2. For each failure calculate the adjusted rank as follows:

$$R_i = R_{i-1} + \frac{n+1-R_{i-1}}{n+2-S_i} \tag{7.19}$$

where, R_i is the adjusted rank of the ith failure, R_{i-1} is the adjusted rank of the $(i\text{-}1)^{th}$ failure, that is the previous failure. R_0 is zero and S_i is the sequence number of the ith failure.

As a quick check, the adjusted rank of the ith failure will always be less than or equal to the sequence number and at least 1 greater than the previous adjusted rank. If there is no suspensions, the adjusted rank will be equal to the sequence number as before. These adjusted ranks are then substituted into the Benard's approximation formula to give the median rank and the estimate for cumulative probability is given by:

$$\hat{F}(t_i) = \frac{R_i - 0.3}{n+0.4}$$

Table 7-5. Estimated failure and reliability function

S_i	t_i	j	$R_j = R_{j-1} + [(n+1- R_{j-1}) / (n+2 - S_i)]$	$\hat{F}(t_i)$	$\hat{R}(t_i)$
1	2041	1	1	0.0565	0.9435
2	2173	2	2	0.1370	0.8630
3	2248*				
4	2271	3	3.1	0.2258	0.7742
5	2567*				
6	2665*				
7	3008	4	4.51	0.3395	0.6605
8	3091	5	5.92	0.4532	0.5468
9	3404*				
10	3424	6	7.69	0.5960	0.4040
11	3490*				
12	3716	7	10.34	0.8097	0.1903

EXAMPLE 7.3

The following data were observed during the data capturing exercise on 12 compressors that are being used by different operators. Estimate the reliability and failure function (* indicates that the data is a censored data)

2041, 2173, 2248*, 2271, 2567*, 2665*, 3008, 3091, 3404*, 3424, 3490*, 3716

We need to calculate the adjusted rank of the failure times using equation (7.19), once this is done, and then the failure and reliability function can be estimated using equations (7.2) and (7.3) respectively. The estimated failure and reliability functions are shown in Table 7.5.

7.4 REGRESSION

The models used to relate a dependent variable y to the independent variables x are called regression models. The simplest regression model is the one that relates a dependent variable y to a single independent variable x in a linear relationship (*linear regression model).* Linear regression provides predicted values for the dependent variables (y) as a linear function of independent variable (x). That is, linear regression finds the best-fit straight line for the set of points (x, y). The objectives of linear regression are:

1. To check whether there is a linear relationship between the dependent variable and the independent variable.
2. To find the best fit straight line for a given set of data points.
3. To estimate the constants 'a' and 'b' of the best fit y = a + bx.

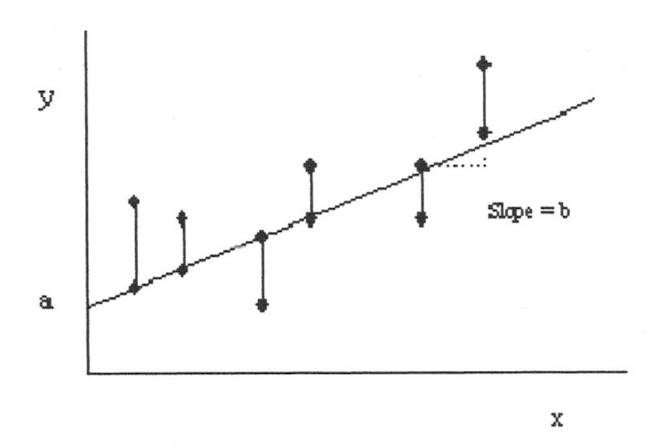

Figure 7-3. Least square regression

The standard method for linear regression analysis (fitting a straight line to a single independent variable) is using the *method of least squares*. Least square regression is a procedure for estimating the coefficients 'a' and 'b' from a set of X, Y points that have been measured. In reliability analysis, the set X is the set of time to failures (or function of TTF) and set Y is their corresponding cumulative probability values (or function of cumulative distribution). In this book, the cumulative probability values are estimated using Median Rank and thus the regression technique is called *Median Rank Regression* (MRR). Figure 7.3 illustrates the least square regression. The measure of how well this line fits the data is given by the correlation coefficient. If we construct a line such that it passes through the point (x^*, y^*) where x^* is the mean of the x values and y^* is the mean of the y values then the sum of the distances between each point and the point on the line vertically above (-ve) or below (+ve) will always be zero (provided the line is not parallel to the y-axis). The same holds for the horizontal distances provided that the line is not parallel to the x-axis. This means that any line passing through the means (in the way described) will be an unbiased estimator of the true line.

If we now assume that there is a linear relationship between the x's (x $\in$ X) and y's (y $\in$ Y), that the x's are known exactly and that the "errors" in the y values are normally distributed with mean 0 then it can be shown that the values of a and b which minimize the expression:

$$\sum_{i=1}^{n}(y_i - a - bx_i)^2 \tag{7.20}$$

will give the best fit. The expression ($y_i - a - bx_i$) gives the vertical distance between the point and the line. Cutting out lot of algebra, one can show that the values of a and b can be found by solving the following two equations:

$$na + b\sum_{i=1}^{n} x_i = \sum_{i=1}^{n} y_i \tag{7.21}$$

$$a\sum_{i=1}^{n} x_i + b\sum_{i=1}^{n} x_i^2 = \sum_{i=1}^{n} x_i y_i \tag{7.22}$$

'a' is the estimate of the intercept (of the line with the y-axis) and 'b' is the estimate of the slope – i.e. y = a + bx is the equation of the line giving:

$$b = \frac{n\sum_{i=1}^{n} x_i y_i - \sum_{i=1}^{n} x_i \sum_{i=1}^{n} y_i}{n\sum_{i=1}^{n} x_i^2 - (\sum_{i=1}^{n} x_i)^2} \tag{7.23}$$

$$a = \sum_{i=1}^{n} \frac{y_i}{n} - b\sum_{i=1}^{n} \frac{x_i}{n} \tag{7.24}$$

Note also that these expressions are not symmetrical in x and y. The formula quoted here gives what is called *"y on x" regression* and it assumes the errors are in the y-values.

By replacing each x with a y and each y with an x we can perform "x on y" regression (which assumes that the errors are in the x-values). If *c* is the estimate of the intercept so obtained and d is the estimate of the slope then to get estimates of *a* and *b* (the intercept and slope of the original graph):

$$b = \frac{1}{d} \text{ and } a = -\frac{c}{d}$$

Note: unless the points are collinear, the "x on y" estimates will not be the same as the "y on x" estimates. In the special case where you want to force the line through the origin (i.e. the intercept is zero), the least squares formula for the slope becomes:

$$b = \frac{\sum_{i=1}^{n} x_i y_i}{\sum_{i=1}^{n} x_i^2} \tag{7.25}$$

7.4.1 Correlation co-efficient

A measure of the dependence between two variables is given by the correlation coefficient. The correlation coefficient, r is given by:

$$r = \frac{n\sum_{i=1}^{n} x_i y_i - \sum_{i=1}^{n} x_i \sum_{i=1}^{n} y_i}{\sqrt{n\sum_{i=1}^{n} x_i^2 - (\sum_{i=1}^{n} x_i)^2} \times \sqrt{n\sum_{i=1}^{n} y_i^2 - (\sum_{i=1}^{n} y_i)^2}} \tag{7.26}$$

The correlation coefficient always lies between -1 and $+1$. A value of $+1$ or -1 means that x and y are exactly linearly related. In the former case y increases as x increases but for r = -1, y decreases as x increases. Note that if x and y are independent then r = 0, but r = 0 does not mean that x and y are independent. The best fit distribution is the one with maximum r value (close to one). To find the best fit, regression analysis is carried out on the popular distribution such as exponential, Weibull, normal and log-normal. The one with highest correlation coefficient is selected as the best. The coordinates (x, y) and the corresponding parameters for different distributions are given in the following sections.

7.4.2 Linear regression for exponential distribution

To fit a data to an exponential distribution, we transform the co-ordinates (t_i, $F(t_i)$) in such a way that, when plotted, it gives a straight line. Here t_i is the observed failure times and $F(t_i)$ is the estimated cumulative distribution function. The cumulative distribution of exponential distribution is given by:

$$F(t) = 1 - \exp(-\lambda t)$$

that is,

$$\ln[\frac{1}{1-F(t)}] = \lambda t \tag{7.27}$$

Equation (7.27) is a linear function. Thus, for an exponential distribution, the plot of $(t, \ln[\frac{1}{1-F(t)}])$ provides a straight line. Thus, if t_1, t_2, ..., t_n are the observed failure times, then to fit this data into an exponential distribution, we set:

$$x_i = t_i \tag{7.28}$$

$$y_i = \ln[\frac{1}{1-F(t_i)}] \tag{7.29}$$

Substituting (x_i , y_i) in equation (7.23) we get:

$$b = \frac{\sum_{i=1}^{n} x_i y_i}{\sum_{i=1}^{n} x_i^2} \tag{7.30}$$

Note that, for exponential distribution b = 1/MTTF.

EXAMPLE 7.4

The following failure data were observed on Actuators. Fit the data to an exponential distribution and find the MTTF and the correlation coefficient.

14, 27, 32, 34, 54, 57, 61, 66, 67, 102, 134, 152, 209, 230

First we carry out least square regression on $t_i, \ln[\frac{1}{1-F(t_i)}]$, various calculations are tabulated in Table 7.6.

Table 7-6. Regression analysis of data in example 7.4

i	t_i (= x_i)	$F(t_i)$	$y_i = ln[1/(1-F(t_i))]$
1	14	0.0486	0.0498
2	27	0.1180	0.1256
3	32	0.1875	0.2076
4	34	0.2569	0.2969
5	54	0.3263	0.3951
6	57	0.3958	0.5039
7	61	0.4652	0.6260
8	66	0.5347	0.7651
9	67	0.6041	0.9267
10	102	0.6736	1.1196
11	134	0.7430	1.3588
12	152	0.8125	1.6739
13	209	0.8819	2.1366
14	230	0.9513	3.0239

The value of b is given by:

$$b = \frac{\sum_{i=1}^{n} x_i y_i}{\sum_{i=1}^{n} x_i^2} = \frac{\sum_{i=1}^{n} t_i \times \ln[\frac{1}{1-F(t_i)}]}{\sum_{i=1}^{n} t_i^2} = 0.01126$$

MTTF is given by 1/b = 1/0.01126 = 88.73. The corresponding correlation coefficient is 0.983 (giving r^2= 0.9666).

7.4.3 Linear regression for Weibull distribution

Cumulative distribution of Weibull distribution is given by:

$$F(t) = 1 - \exp(-(\frac{t}{\eta})^{\beta})$$

That is, $\ln[\ln(\frac{1}{1-F(t)})] = \beta \ln(t) - \beta \ln(\eta)$, which is a linear function. Thus to fit the data to a Weibull distribution, we set:

$$x_i = \ln(t_i) \tag{7.31}$$

$$y_i = \ln[\ln(\frac{1}{1-F(t_i)})] \tag{7.32}$$

From least square regression, it is evident that the shape and scale parameters of the distribution are given by:

$$\beta = b \tag{7.34}$$

$$\eta = \exp(-a / \beta) \tag{7.35}$$

Table 7-7. Weibull regression of data in example 7.5

i	t_i	$F(t_i)$	$x_i = ln(t_i)$	$Y_i = lnln(1/1\text{-}F(t_i))$
1	17	0.0614	2.8332	- 2.7581
2	21	0.1491	3.0445	- 1.8233
3	33	0.2368	3.4965	- 1.3082
4	37	0.3245	3.6109	- 0.9354
5	39	0.4122	3.6635	- 0.6320
6	42	0.5	3.7376	- 0.3665
7	56	0.5877	4.0253	- 0.1209
8	98	0.6754	4.5849	0.1180
9	129	0.7631	4.8598	0.3648
10	132	0.8508	4.8828	0.6434
11	140	0.9385	4.9416	1.0261

EXAMPLE 7.5

Construct a least square regression for the following failure data:

17, 21, 33, 37, 39, 42, 56, 98, 129, 132, 140

Making use of equations (7.31) and (7.32), we construct the least square regression, as described in Table 7.7.

Using equations (7.34) and (7.35), we get $\beta = 1.4355$, $\eta = 76.54$ and the correlation coefficient r = 0.9557 ($r^2 = 0.9133$).

7.4.4 Linear regression for normal distribution

For normal distribution,

$$F(t) = \Phi(\frac{t-\mu}{\sigma}) = \Phi(z)$$

z can be written as:

$$z_i = \Phi^{-1}[F(t)] = \frac{t_i - \mu}{\sigma} = \frac{t_i}{\sigma} - \frac{\mu}{\sigma} \tag{7.36}$$

Equation (7.36) is a linear function. For regression, we set $x_i = t_i$ and $y_i = z_i$ ($= \Phi^{-1}$ $[F(t_i)]$). The value of z can be obtained from standard normal distribution table. One can also use the following expression that gives polynomial approximation for z_i.

$$x_i = t_i \tag{7.37}$$

$$y_i = P - \frac{C_0 + C_1 P + C_2 P^2}{1 + d_1 P + d_2 P^2 + d_3 P^3} \tag{7.38}$$

where

$$P = \sqrt{\ln[\frac{1}{[1 - F(t_i)]^2}]} \tag{7.39}$$

$C_0 = 2.515517$, $C_1 = 0.802853$, $C_2 = 0.010328$, $d_1 = 1.432788$,
$d_2 = 0.189269$, $d_3 = 0.001308$

The estimates for μ and σ are given by

$$\mu = -\frac{a}{b} \text{ and } \sigma = \frac{1}{b} \tag{7.40}$$

EXAMPLE 7.6

Fit the following data into a normal distribution

62, 75, 93, 112, 137, 170, 185

Table 7.8 gives various computations involved in regression.

Table 7-8. Regression of data shown in example 7.6

i	t_i	$F(t_i)$	$z_i = P - (c_0 + c_1 P + c_2 P^2 / 1 + d_1 P + d_2 P^2 + d_3 P^3)$
1	62	0.0945	- 1.2693
2	75	0.2297	- 0.7302
3	93	0.3648	- 0.3434
4	112	0.5	0
5	137	0.6351	0.3450
6	170	0.7702	0.7394
7	185	0.9054	1.3132

7.5 MAXIMUM LIKELIHOOD ESTIMATION (MLE)

Although there are many benefits in using regression methods it should be recognized that there are other methods. One of these, which can be useful in many circumstances, is *maximum likelihood estimation (MLE).*

Although regression is widely used for parameter estimation, it requires strong assumptions like linearity between dependent and independent variables or knowledge of distribution and non-correlation of independent variables with the error term etc. on the sample data. In cases where these assumptions are not satisfied, one has to employ other methods like parametric and non-parametric methods that do not require such strong assumptions. One of the next most popular parametric methods that does not need a linearity assumption, but requires the knowledge of distribution is *maximum likelihood estimation* (MLE).

The main advantage of MLE is that it is mathematically more rigorous and is less susceptible to individual values as every time-to-failure has equal weighting. Its disadvantages are that it is more difficult to use, it does not provide a visual check and its point estimates of the shape (β) parameter tends to be biased for small samples.

The method can be used with any type of distribution. It can also be used with data that is censored or in which the failures are only known to have occurred at sometime during a given (time) interval.

7.5.1 Complete and uncensored data

Consider a scenario in which n identical items are operated until they have all failed or, equivalently, one item is operated until it has failed n times with each repair restoring the item to an "as-good-as-new" condition and its age being reset to zero. Let us further assume that the times at which each of these items failed are known to be $t_1, t_2, \ldots, t_n$. If we now assume that these are all independent and identically distributed (*iid*) with probability density function $f(t : \overline{\theta})$ where $\overline{\theta}$ is the set of parameters, then we can calculate the likelihood:

$$l = \prod_{i=1}^{n} f(t_i : \vec{\theta}) \tag{7.41}$$

Strictly speaking, the product should be multiplied by n! as the order of the times is unimportant so that we would say the sequence $t_n, t_{n-1}, \ldots, 2, 1$ is the same as the one given and is, clearly, equally likely. However, because we are not interested in the actual "probability" of the given scenario, only in what values of the parameters maximize the "likelihood", we can safely ignore the n!

To make mathematics easier and to reduce the problems of dealing with numbers very close to zero, it is a normal practice to use a (natural) logarithm transform:

$$-\log_e(l) = -\ln(l) = L = -\sum_{i=1}^{n} \ln(f(t_i : \vec{\theta})) \tag{7.42}$$

Note: Because *f(t)* will always be between 0 and 1, the (natural) log will always be negative. In order to make it a positive value, add the negative sign and turn it into a minimization.

Of course, in general, we do not know the true values of the population parameters. The maximum likelihood estimation method is based on the assumption that if we can find values for the parameters that maximize the likelihood (i.e. minimizes the value of L) then these should be the "best" estimates of the true values.

Now, the maxima and minima of any function, L, occur at:

$$\frac{dL}{d\vec{\theta}} = 0 \tag{7.43}$$

So all we have to do is find the derivatives of L with respect to each of the parameters, substitute in the values of the TTF's and solve the simultaneous equations to obtain the "best" estimates of the parameters. Note that at this stage we have not specified any particular distribution, only that all of the TTF's are iid (independent and identically distribution, i.e. from the same distribution) so we could actually extend this one stage further by choosing the type of distribution which gives the lowest value of L_{mle} (i.e. maximises the maximum likelihood estimates). However, before we do that, let us consider some of the more common distributions.

7.5.2 Maximum likelihood estimator of exponential distribution

The exponential distribution only has one parameter which can either be specified as the mean time to failure (MTTF) ($1/\lambda$) or as its reciprocal, the failure rate, λ. The probability density function is given by:

$$f(t) = \lambda e^{-\lambda t}$$

The expression for the likelihood function becomes:

$$l = \prod_{i=1}^{n} f(t_i) = \prod_{i=1}^{n} \lambda e^{\lambda t_i} \tag{7.44}$$

or, if we consider the negative log-likelihood, this becomes:

$$L = \sum_{i=1}^{n} \{-\ln(f(t_i))\} = \sum_{i=1}^{n} \{\lambda t_i - \ln(\lambda)\} \tag{7.45}$$

giving

$$L = \lambda \sum_{i=1}^{n} t_i - n \ln(\lambda) \tag{7.46}$$

Now, the minimum value of L will occur when

$$\frac{\partial L}{\partial \gamma} = 0$$

that is:

$$\frac{\partial L}{\partial \gamma} = \frac{\partial}{\partial \gamma}(\lambda \sum_{i=1}^{n} t_i - n \ln(\lambda)) = \sum_{i=1}^{n} t_i - \frac{n}{\lambda} = 0 \tag{7.47}$$

or

$$\frac{n}{\lambda} = \sum_{i=1}^{n} t_i$$

giving

$$\frac{1}{\lambda} = \frac{1}{n} \sum_{i=1}^{n} t_i = \bar{t} = \hat{\text{MTTF}} \tag{7.48}$$

Where, $\bar{t}$ is the arithmetic mean of times to failure. Thus the mean time to failure is, in fact, the maximum likelihood estimator of the parameter of the exponential distribution.

EXAMPLE 7.7

Suppose it has been decided to demonstrate the reliability of a turbine disc by running a number of discs on spin rigs until they burst. The following times were recorded as the times-to-failure for each of the 5 discs: 10496, 11701, 7137, 7697 and 7720 respectively.

Making the assumption that these times are exponentially distributed then we can find the MLE of the parameter as

$$\hat{MTTF} = \frac{1}{n}\sum_{1}^{n} t_i = \frac{10496 + 11701 + 7137 + 7697 + 7720}{5} = 8950.2$$

We can also determine the value of L from equation (7.46) as:

$$L = \gamma \sum_{i=1}^{n} t_i - n\ln(\gamma) = 5 - (-45.497) = 50.497$$

Note: this does not mean that the times are exponentially distributed and it does not say anything about how well the data fits this distribution; for that we will need to look at interval estimators.

7.5.3 Maximum likelihood estimator for Weibull distribution

The Weibull distribution has two parameters: β and η. Its probability density function is given by:

$$f(t) = \frac{\beta t^{\beta-1}}{\eta^{\beta}} e^{-\left(\frac{t}{\eta}\right)^{\beta}} = \beta \psi t^{\beta-1} e^{-\psi t^{\beta}} \quad \text{where } \psi = \eta^{-\beta}$$

The expression for the likelihood function becomes:

$$l = \prod_{i=1}^{n} f(t_i) = \prod_{i=1}^{n} \beta \psi t_i^{\beta-1} e^{-\psi t_i^{\beta}} \tag{7.49}$$

and

$$L = -\sum_{i=1}^{n} \ln(\beta \psi t_i^{\beta-1} e^{-\psi t_i^{\beta}}) \tag{7.50}$$

giving,

$$L = -n\ln(\beta) - (\beta - 1)\sum_{i=1}^{n} \ln(t_i) - n\ln(\psi) + \psi \sum_{i=1}^{n} t_i^{\beta} \tag{7.51}$$

Now, the minimum is given when

$$\frac{\partial L}{\partial \vec{\theta}} = 0 \text{ i.e. } \frac{\partial L}{\partial \beta} = \frac{\partial L}{\partial \psi} = 0 \text{, giving}$$

$$\frac{\partial L}{\partial \beta} = -\frac{n}{\beta} - \sum_{i=1}^{n} \ln(t_i) + \psi \sum_{i=1}^{n} t_i^{\beta} \ln(t_i) = 0$$

$$\frac{\partial L}{\partial \psi} = -\frac{n}{\psi} + \sum_{i=1}^{n} t_i^{\beta} = 0$$

Taking the second equation first gives:

$$\frac{1}{\psi} = \frac{1}{n} \sum_{i=1}^{n} t_i^{\beta} \quad \text{or } \lambda = \left(\frac{1}{n} \sum_{i=1}^{n} t_i^{\beta} \right)^{1/\beta} \tag{7.52}$$

We can now substitute this expression for ψ into the first differential equation to give:

$$\frac{n}{\beta} + \sum_{i=1}^{n} \ln(t_i) - \frac{n \sum_{i=1}^{n} t_i^{\beta} \ln(t_i)}{\sum_{i=1}^{n} t_i^{\beta}} = 0 \tag{7.53}$$

This is now independent of ψ so is relatively easy to solve numerically using Newton-Raphson or similar search method. In MicroSoft™ Excel® Solver® can be used to solve this expression or, one could actually solve the original likelihood expression. However, one has to be careful in setting the precision and conditions, recognizing that neither β nor ψ (η) can take negative values. One may need to scale the times (by dividing them all by the mean or 10^k, say) to avoid overflow problems particularly if β may become "large" (> 10, say) which can occur if n is small (<10).

EXAMPLE 7.8

Using the same data as above in the Exponential case, we can now find, using Solver®, the value of β that satisfies equation 7.53. The computations are shown in Table 7.9.

Table 7-9. Calculation of Weibull MLE

	Times	Ln(T)	T^{β}	$Ln(T)*T^{\beta}$
T_1	10496	9.25875	3.18E+21	2.9457E+22
T_2	11701	9.36743	5.69E+21	5.32914E+22
T_3	7137	8.873048	4.04E+20	3.58881E+21
T_4	7697	8.948586	6.06E+20	5.4208E+21
T_5	7720	8.95157	6.16E+20	5.50982E+21
Sums	44751	45.39938	1.05E+22	9.72678E+22

In Table 7.9, the value for β is 5.3475...

If we substitute the values for β and those in Table 7.9 into equation 7.53 we get:

$$\frac{5}{5.3475} + 45.39938 - \frac{5 * 9.72678 * 10^{22}}{1.05 * 10^{22}} = 6.45237 * 10^{-10}$$

Note: The right-hand side is not equal to zero but is within the tolerances set in Solver®.

By using the same values we can calculate L = 44.720.

This value is less than that obtained for the Exponential (= 50.497) which suggests that the Weibull is a better fit, thus suggesting the times to failure are age-related. Unfortunately, it does not tell us how good the fit is or, indeed, if another (theoretical) distribution might fit the data more closely.

7.5.4 Maximum likelihood estimator for normal distribution

The normal distribution differs from the previous two distributions in so far as it is defined for both positive and negative values but this does not affect the MLE process. The probability density function for the normal distribution is given by:

$$f(x:\mu,\sigma)=\frac{1}{\sigma\sqrt{2\pi}}e^{-\frac{(x-\mu)^2}{2\sigma^2}}$$

Giving

$$l=\prod_{i=1}^{n}\frac{1}{\sigma\sqrt{2\pi}}e^{-\frac{(x_i-\mu)^2}{2\sigma^2}} \tag{7.54}$$

and

$$L=\sum_{i=1}^{n}\left\{\frac{1}{2}\ln(2\pi)+\ln(\sigma)+\frac{(x_i-\mu)^2}{2\sigma^2}\right\} \tag{7.55}$$

The maxima and minima of L occur when

$$\frac{\partial L}{\partial \mu}=\frac{\partial L}{\partial \sigma}=0$$

$$\frac{\partial L}{\partial \mu}=\frac{1}{2\sigma^2}\sum_{i=1}^{n}-2(x_i-\mu)=0$$

which, can be reduced to

$$\mu=\frac{1}{n}\sum_{i=1}^{n}t_i=\bar{t} \tag{7.56}$$

That is, the maximum likelihood estimator of the mean is simply the sample mean.

$$\frac{\partial L}{\partial \sigma} = \frac{n}{\sigma} - \frac{1}{\sigma^3}\sum_{i=1}^{n}(x_i - \mu)^2$$

Which can be reduced to:

$$\sigma^2 = \frac{1}{n}\sum_{i=1}^{n}(x_i - \mu)^2 \tag{7.57}$$

σ^2 is the definition of the (population) variance.

EXAMPLE 7.9

Returning to our 5 discs, we can now apply these two formulae to determine the MLE of the mean and variance.

Solution:

Table 7.10 shows the calculations involved in MLE of normal distribution

Table 7-10. MLE calculations for normal distribution

Discs	Times	$(x-\mu)^2$
T_1	10496	2389498
T_2	11701	7566901
T_3	7137	3287694
T_4	7697	1570510
T_5	7720	1513392
Mean & Var.	8950.2	3265599
St. Dev.		1807.097

This gives μ_{MLE} = 8950.2 and σ_{MLE} = 1807.1. Substituting these point estimates of μ and σ into equation (7.55), we get the log-likelihood

L = 44.59

This value is slightly lower than that obtained for the Weibull (44.72) which suggests that the normal distribution provides a (marginally) better fit to the data than the Weibull. This, again, indicates that the cause of failure is age-related.

7.6 CASE STUDY: ENGINEER TANK

In this section we continue with the engineer tank case discussed in Chapter 3. Table 7.11 lists faults found during inspection and groups the faults under generic headings (it also includes the time to failure data given in Table 3.10 and illustrative engine numbers). For example, if during an engine overhaul, signs of wear were identified on the valve guides, bores and piston rings and the oil was very dirty, the generic cause of failure can be considered to be dirt contamination.

The data in Table 7.11 can be used to identify major failure modes and failure causes. Pareto charts detailing both the failure and generic failure modes are shown in Figures 7.4 and 7.5 respectively

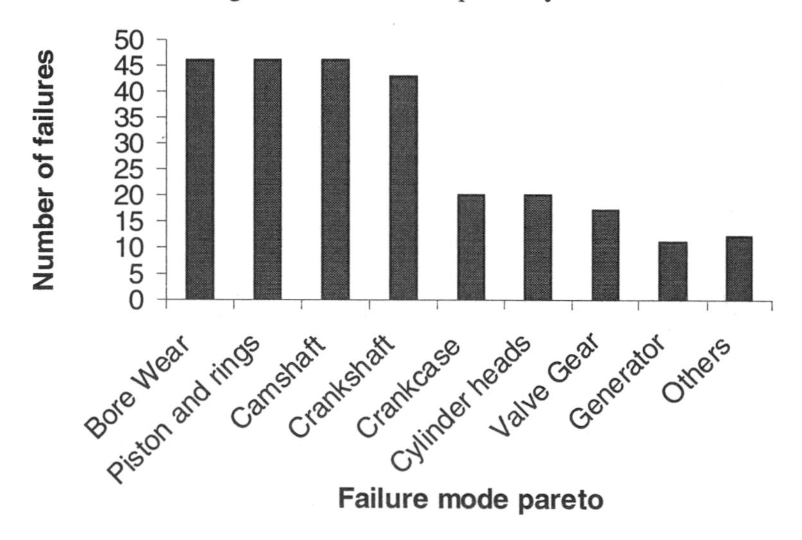

Figure 7-4. Generic failure modes in engineer tank

Failure Cause Pareto shows that dust ingestion is a major cause of failure. This is not unusual for armoured vehicles which work in an environment hostile to air filters

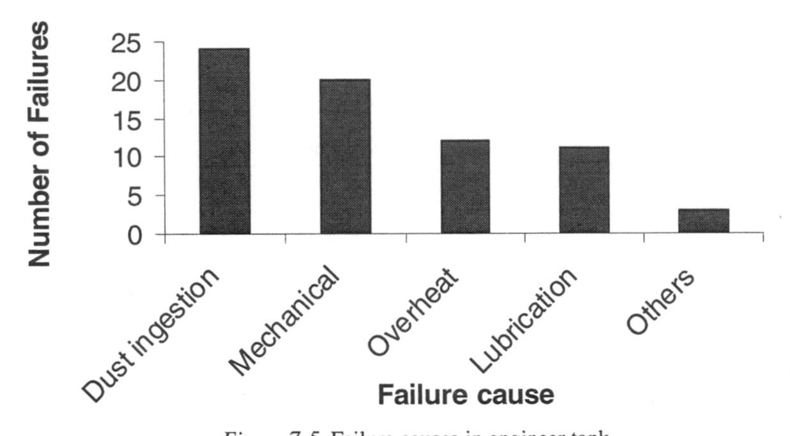

Figure 7-5. Failure causes in engineer tank

Regression is preformed to the failure data in Table 3.10 to identify the best distribution. The regression results are shown in Figure 7.6. The reliability function and hazard function based on the best fit Weibull distribution are shown in Figure 7.7.

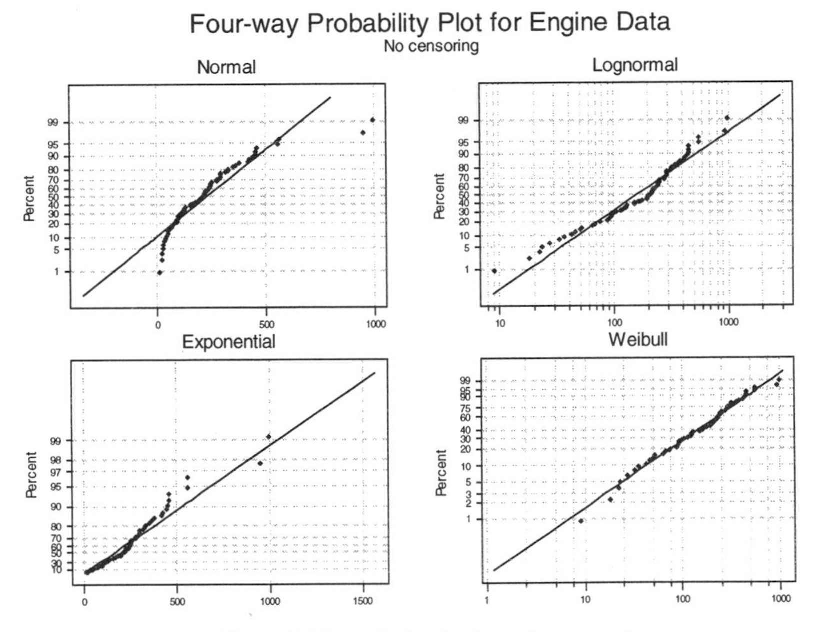

Figure 7-6. Best fit distribution using regression

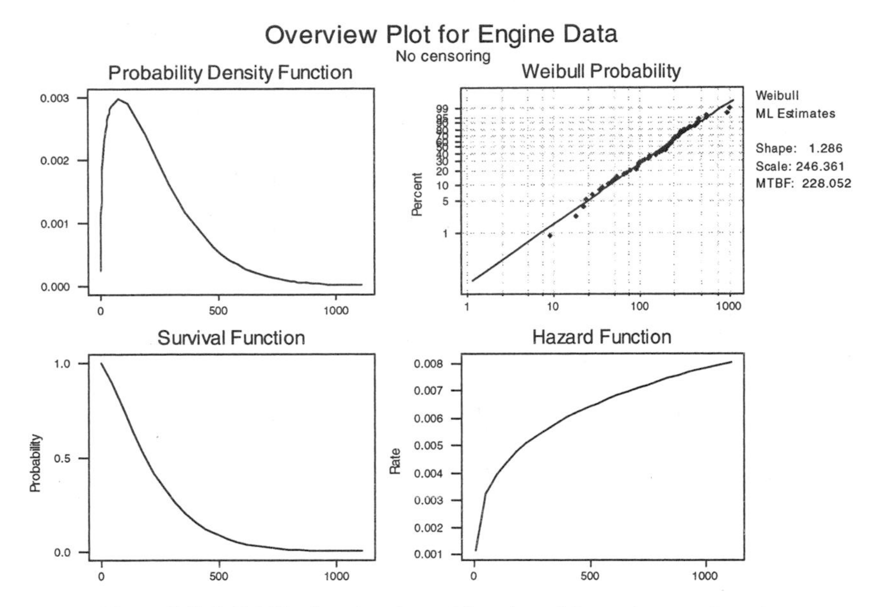

Figure 7-7. Reliability function, hazard function of the engineer tank engine

The engine MTBF is 228 hours against the target 420. The failure mode and failure cause will help to improve the reliability of the engine by providing important information about the design weaknesses. For example, from Figure 7.5, it is clear that dust ingestion is one of the main causes of failure and the design should focus to design better filters (self cleaning filters) that reduce the engine failures due to dust ingestion.

Table 7-11. Engine strip data

Strip Report Physical Fault													Component Generic Fault				
Engine No	Time To Failure	Bore Wear	Pistons and rings	FIE Failure	Valve gear	Crankshaft or bearings	No FF	Camshaft fault	Gen drive	Crank case	Fans	Cylinder Heads	Lubrication	Overheat	Mechanical	Dust Ingestion	No Fault Found
R4	215	0	0	0	1	1	0	1	1	0	0	0	0	0	1	0	0
L2	114	0	0	0	0	0	0	0	0	1	0	0	0	0	1	0	0
M7	247	1	1	0	0	1	0	0	0	0	0	0	0	0	0	1	0
F9	122	1	1	0	0	0	0	0	0	0	0	0	0	0	0	1	0
M5	91	1	1	0	0	0	0	0	0	0	0	0	0	0	0	1	0
L4	291	1	1	0	0	0	0	0	0	0	0	0	0	0	0	1	0
I5	194	0	0	0	0	0	1	0	0	0	0	0	0	0	0	0	1
J8	315	1	1	0	1	1	0	0	0	0	0	0	0	0	0	1	0
W2	241	0	0	0	0	0	0	0	0	1	0	0	1	0	0	0	0
Q7	9	1	1	0	0	1	0	1	0	0	0	0	1	0	0	0	0
W4	178	0	1	0	0	0	0	0	0	0	0	1	0	1	0	0	0
E2	33	1	1	0	0	1	0	0	0	1	0	0	1	0	0	0	0
R3	997	1	1	0	1	1	0	0	0	0	0	0	0	0	1	0	0
T1	75	0	0	0	1	0	0	1	0	0	0	1	0	1	0	0	0
Y1	95	0	0	0	1	1	0	0	1	0	0	0	0	0	1	0	0

Strip Report Physical Fault													Component Generic Fault				
I2	378	1	1	0	0	1	0	0	0	0	0	0	0	0	0	1	0
O2	425	0	1	0	0	1	0	0	0	1	0	0	1	0	0	0	0
P5	252	1	1	0	0	0	0	0	0	0	0	0	0	0	0	1	0
A9	22	1	1	0	1	0	0	0	0	0	0	1	0	0	0	1	0
S7	153	1	1	0	1	1	0	0	0	1	0	1	0	1	0	0	0
D4	195	1	1	0	1	0	0	0	0	0	0	1	0	0	0	1	0
F8	165	1	1	0	0	0	0	0	0	0	0	0	0	0	0	1	0
G9	130	0	0	0	0	0	0	0	0	1	0	0	0	0	1	0	0
H2	451	1	1	0	0	1	0	0	0	0	0	0	0	0	0	1	0
J3	126	1	1	0	0	0	0	0	0	0	0	1	0	1	0	0	0
K5	23	1	1	0	1	1	0	1	0	0	0	1	0	0	1	0	0
L5	416	0	0	0	0	0	0	0	0	1	0	0	0	0	1	0	0
Z7	247	0	0	0	0	0	0	1	0	1	0	0	0	1	0	0	0
X9	456	1	1	0	0	1	0	0	0	1	0	0	0	0	0	1	0
C8	350	1	1	0	0	1	0	0	0	0	0	0	0	0	0	1	0
V6	105	1	1	0	1	1	0	1	0	0	0	0	0	1	0	0	0
B2	275	1	1	0	1	1	0	0	1	0	0	1	0	0	0	1	0
N1	294	1	1	0	0	1	0	0	0	0	0	0	0	0	0	1	0
M4	232	1	1	0	0	1	0	0	0	1	0	0	1	0	0	0	0
U6	441	0	0	0	0	0	1	0	0	0	0	0	0	0	0	0	1
A1	94	1	1	1	0	1	0	0	0	0	0	0	0	0	1	0	0

Strip Report Physical Fault													Component Generic Fault				
S2	360	0	0	0	0	0	0	0	0	1	0	0	1	0	0	0	0
D4	202	0	0	1	1	0	0	1	0	0	0	0	0	0	1	0	0
F3	210	0	0	0	0	1	0	0	0	0	1	1	0	1	0	0	0
G5	36	1	0	0	0	1	0	0	0	1	0	1	1	0	0	0	0
H6	46	0	0	1	0	1	0	0	1	0	0	0	0	0	1	0	0
J8	223	1	1	0	0	0	0	0	0	0	0	1	0	1	0	0	0
K7	18	1	1	1	0	1	0	1	0	0	0	0	0	0	1	0	0
L9	50	1	1	0	0	0	0	0	0	1	0	0	0	0	0	1	0
Q1	68	1	1	0	1	1	0	0	1	0	0	1	0	0	1	0	0
W2	244	0	0	0	0	1	0	0	1	0	0	0	0	0	1	0	0
E3	126	1	1	1	1	0	0	0	0	0	0	1	0	0	0	1	0
R7	557	0	0	0	0	1	0	0	0	1	0	0	1	0	0	0	0
T4	168	1	1	0	1	1	0	0	1	0	0	0	0	0	1	0	0
Y5	64	1	1	0	0	1	0	0	0	1	0	0	0	0	1	0	0
U6	99	0	0	0	0	1	0	0	1	0	1	1	0	1	0	0	0
I7	323	1	1	0	0	1	0	0	0	0	0	1	0	0	0	1	0
O8	284	0	0	0	0	1	0	0	1	0	0	0	0	0	1	0	0
P9	52	1	1	0	0	1	0	0	0	0	0	0	1	0	0	0	0
Q2	223	1	1	0	0	0	0	0	0	1	1	1	0	1	0	0	0
Z3	454	0	0	0	0	1	0	0	0	0	0	1	1	0	0	0	0
Y4	42	1	1	0	0	1	0	0	0	0	0	0	0	0	0	1	0

Strip Report Physical Fault													Component Generic Fault				
B5	560	1	1	0	0	1	0	0	0	1	0	0	1	0	0	0	0
D6	232	1	1	0	0	1	0	0	0	0	0	1	0	1	0	0	0
S7	198	0	0	0	0	1	0	0	0	0	1	1	0	1	0	0	0
L8	27	1	1	0	0	1	0	0	0	0	0	0	0	0	0	1	0
O9	947	1	1	0	0	1	0	0	0	0	0	0	0	0	0	1	0
J0	239	1	0	0	0	0	1	0	0	0	0	0	0	0	0	0	1
K1	89	0	0	0	0	0	0	0	1	0	0	0	0	0	1	0	0
N9	325	1	1	0	0	1	0	0	0	1	0	0	0	0	1	0	0
H8	273	1	1	0	1	1	0	1	0	0	0	1	0	0	0	1	0
M7	292	1	1	0	0	0	0	0	0	0	0	0	0	0	0	1	0
N6	216	0	0	0	1	0	0	0	0	0	0	0	0	0	1	0	0
D5	151	1	1	0	0	1	0	0	1	1	0	0	0	0	1	0	0
S4	88	1	1	0	0	1	0	0	0	1	0	0	0	0	0	1	0
Total		46	46	5	17	43	3	46	11	20	4	20	11	12	20	24	3

Chapter 8

SOFTWARE RELIABILITY

To err is human - and to blame it on a computer is even more so.
Robert Orben

8.1 INTRODUCTION

Some of the most complex systems Man has built are in software. In recent years many hardware functions have been replaced by software and thus software has become a critical part of many systems. Starting from fighter aircraft to household electronic items software plays a major role. Typical examples from the most common applications include the software controlled fuel injection system of the automobile [whose performance is vastly superior to the best carburetor] and the automobile braking system incorporating anti-skid and traction control functions. Software is the core element of today's automobile and medical industries. A growing proportion of the systems operate in real time. The operational effects of failure are large and critical. For example a breakdown of airline reservations, train reservations, banking and many other services can bring the day to day activities to a stand still. The failure of a software controlled braking system in an automobile can be disastrous. For systems such as air traffic control, space shuttle, fighter aircraft and automated missiles, the software makes a very crucial contribution to the success of the mission. Software cannot be seen or touched, but it is essential to applications such as those described above. It is necessary that the reliability of software be measured and evaluated, as it is in hardware. It is also essential to understand how Six Sigma concepts can be used in software development to ensure quality and reliability.

There are fundamental differences between the methods used for the reliability analysis of software and hardware. The design fault in software is the main cause of its failure whereas physical deterioration causes hardware failure. Software generally becomes more reliable over time because of successive changes made with experience (however it might become obsolete as the technology changes). Failure of software is a departure of its results from the requirement. A failure occurs when the user perceives that the program ceases to deliver the expected service. The terms errors, faults and failures are often used interchangeably, but do have different meanings. In software, an error is usually a programmer action or omission that results in a fault. A fault, also referred to as bug, is a software defect that causes a failure, and a failure is the unacceptable departure of a program operation from the program requirements. When measuring reliability, we are usually measuring only defects found and defects fixed. If the objective is to fully measure reliability, we need to address prevention as well as investigate the development, starting from the requirements phase to the finally developed programs.

Errors in the software are introduced during various stages, mainly during:

- Requirements definition
- Design
- Program development
- Operation /maintenance

Thus any measure of software reliability must start with the core of the issue, operational software error counts and the rate at which they occur; that is the software failure rate. (Koss 1998)

Software failures are not caused by physical environmental wear. The failures occur without any warning. The main source of software failure comes from requirement and specification error rather than from machine code error or design error. Unlike hardware reliability software reliability cannot be improved by redundancy (i.e. producing identical code). However, it is possible to provide redundancy by producing independent versions of software to improve the reliability, which is the basis for many fault tolerant architectures. The software reliability is defined with respect to time and it is generally with respect to execution time. The execution time for a software system is the CPU time that is actually spent by the computer in executing the software.

8.2 SOFTWARE RELIABILITY METRICS

In order to build reliable software the focus must be on comprehensive requirements and a comprehensive testing plan, ensuring all requirements are tested. Importance also should be given for the maintenance of software since there will be a "useful life" phase where sustaining the engineering effort will be needed. Therefore to prevent the errors in the software we must:

- Start with the requirements; ensure that the software is developed in accordance with the requirement specifications.
- Ensure that the code developed can easily support the engineering efforts without infusing additional errors.
- Plan a good comprehensive test program to verify all functionalities stated in the requirement specifications

8.2.1 Requirement reliability metrics

Requirements form the basis for software design, code development and test program. It is critical that the requirements are written such that they are clear and there is no misinterpretation between the developer and the client. For high reliability software, the requirements must be structured, complete and easy to apply.

There are standards like IEEE, DOD and NASA, which give some guidelines for writing the requirements in a structured way. There are several aids to evaluate the quality of the written requirement document. These automated requirement measurements scan the text through the lines and phrases and look for specific words and phrases. For example the weak phrases like 'shall', 'may' which lead to ambiguity will be identified for their frequency of occurrences. More of such terms lead to more choices for the developer to use his discretion.

The Automated Requirements Measurement (ARM) software developed by SATC (Software Assurance Technology Centre), NASA is one such aid for evaluating the quality of the requirement document. Seven measures are taken for determining the quality of the requirement specification document. They are:

1. Lines of text measure the size of the document.
2. Imperatives- words and phrases that ascertain the requirements. The number of imperatives is used as a base requirements count.
3. Continuances-phrases that follow an imperative and introduce the specifications or requirements at a lower level, for a supplemental requirement count.

4. Directives-requirements supported in terms of figures, tables and notes.
5. Weak phrases- words and phrases which are bound to cause uncertainty and leave room for multiple interpretation measure of ambiguity.
6. Incomplete statements- statements that have TBD (to be defined) or TBS (to be supplied).
7. Options-words that give the developer latitude in satisfying the specifications but can be ambiguous.

A tool alone cannot certainly assess the correctness of the requirements specified in the document. However it can assess how well the document is written. For example the count on 5, 6 and 7 clearly would tell how much the developed software can deviate from the requirements the customer bears in his mind as the need. Obviously the acceptance test plan and test will be based on what is contained in the requirement document.

The tool is described as an aid to evaluate the document. However, one can read the document and have several reviews with the customer so that the ambiguity is brought to zero.

8.2.2 Design and Code reliability metrics

The code can be analysed for the structure and architecture to identify possible error-prone modules based on complexity, size and modularity. In general more complex modules are more difficult to understand and have higher probability of defects than less complex modules. Thus the complexity has a direct impact on overall quality and specifically on maintainability. Hence there should be a way to measure the complexity. One way to compute the complexity is to count the number of linearly independent test paths. The size of the software is the count on the number of executable lines of code as defined by a language dependent delimiter.

The combination of size and complexity can be used to evaluate the design and code reliability. Obviously the modules with higher size and complexity are prone to be less reliable. Modules with small size and high complexity are also at a reliability risk because they tend to be very short code that is difficult to change or modify. Although these metrics can be applied to object oriented code, additional metrics are used by SATC (Rosenberg *et al* 2005).

8.2.3 Testing reliability metrics

Testing of reliability metrics can be divided into two parts. The first part is to evaluate the test plan. Testing is the main factor which determines the software reliability. The test simulation for various real world application environments should be planned. Thus it is essential that we evaluate the test plan so that it caters to the testing of all the functionalities specified in the requirements. We have to ensure that each requirement is tested at least once. This should really reduce the number of errors due to lack of expected functionality.

The second part is the evaluation of number of errors and the rate at which they occur. Application of estimation techniques discussed in the next section can be applied. The time to failure data collected can be used to evaluate the reliability of the software. The testing and fixing should be repeated till we attain a satisfactory reliability. In the process one may have to modify the test plans also. Thus the two parts go hand in hand iteratively. The reliability calculated after every corrective action can be used to measure the reliability growth of the software which ultimately tells us the level of maturity of the software development.

8.3 SOFTWARE RELIABILITY MODELS

Software reliability can be defined as follows:

> Software reliability is the probability that software will provide failure free operation in a fixed environment for a fixed interval of time. Probability of failure is the probability that the software will fail on the next selected input.

Software reliability is typically measured per units of time, whereas probability of failure is generally time independent. These two measures can be easily related if one knows the frequency of inputs that are executed per unit of time. Here failure is caused by activation of internal fault or bug. One can view software as a function, which maps a space of inputs into a space of outputs. The input space can be partitioned in to two mutually exclusive sets U and D. Inputs from the set U produce desirable output, whereas inputs from the set D produce incorrect or undesirable output. The software reliability is dependent on the elimination of the input space D to the extent possible by rigorous testing.

Assume that p represents the probability that the software fails for an input selection. The reliability of the software for n inputs selection is given by:

$$R_n = (1-p)^n \tag{8.1}$$

In practice equation (8.1) is not suitable, since the prediction of p is almost impossible. The primary metric for software is the software failure rate, which is given by the expected number of failures per unit time. Here the time refers to a computer execution time. For example if the software fails 5 times during 2000 hours of operation then the failure rate of the software is 0.004. Mathematically,

$$\lambda(t) = \frac{n}{T} \tag{8.2}$$

Where n is the number of failures during **T** hours of execution. Usually the software reliability is measured as number of faults per 1000 lines of code and represented by KLOC. A research study by Pham and Zhang (1999) has shown that professional programmers average six software faults for every 1000 lines of code (LOC) written. At this rate a standard commercial software application of 350,000 LOC will contain over 2000 programming errors. Several internet sites claim that the Windows 2000 had 65,000 bugs when it was released. Windows 2000 has around 35 million lines of source code, in Six Sigma terms; the DPMO for Windows 2000 was around 1857 at the time of release. As the size of the software increases the number of faults increases geometrically. If KLOC is known, then software DPMO can be written as:

$$\text{Software DPMO} = \text{KLOC} \times 10^3 \tag{8.3}$$

Defect in this case is the number of faults per line of source code. The corresponding sigma level is given by:

$$\text{Sigma level for software} = 0.8406 + \sqrt{29.37 - 2.221 \times \ln(\text{KLOC} \times 10^3)} \tag{8.4}$$

The general type of software reliability modeling is divided into two major categories, **prediction** and **estimation.**

8.4 SOFTWARE RELIABILITY PREDICTION MODELS

Prediction models are based on historical models. These are made prior to development or test phases. These can be used as early as the concept phases. These are helpful in predicting reliability in some future time. A few of the prediction models are listed below:

- In-house historical Data collection model
- MUSA's Execution Time Model
- Putnam's Model

All these models are based on a certain estimate of the initial number of faults in the program.

8.4.1 In-house historical data collection model

This model is based on the data collected or database of information accumulated on the software failures during earlier projects. These include failures from the requirement definition stage to the delivery of the software and also the maintenance phase of the software product. Typically statistical regression analysis is used to develop a prediction equation for each of the important project characteristics. This information can be used to predict the reliability of the proposed software product as well as to plan resource allocation.

8.4.2 Musa's execution time model

The model was developed by John Musa of Bell laboratories in the mid 1970's and this was one of the earliest software reliability prediction models. The model predicts the initial failure rate (intensity) of a software system at the point when software testing begins (i.e. when n = 0 and t = 0). The initial failure intensity parameter λ_0 (failures per unit time) is a function of unknown, but estimated, total number of failures N expected in infinite time. The prediction equation is given by (Musa 1998)

$$\lambda_0 = fkw_0 \tag{8.5}$$

The quantity f is the linear execution frequency of the program. This is the average instruction execution rate r divided by the number of object instructions I in the program,

$$f = \frac{r}{I} \tag{8.6}$$

The linear execution frequency is the number of times the program would be executed per unit time if it had no branches or loops. Then it would simply execute all instructions in a linear sequence. The number of object instructions can be estimated by

$$I = \frac{SLOC}{ER} \tag{8.7}$$

SLOC is the source lines of code (not including) reused code. ER is the expansion ratio a constant dependent upon programming language. [For example, assembler, 1.0; Macro Assembler, 1.5; C, 2.5; COBAL, FORTRAN , 3.0; Ada, 4.5; [RAC review (2001)]]. 'k' is a constant which relates failure intensity to "fault velocity". The fault velocity (fw_o) is the average rate at which faults in the program would pass by if the program is executed linearly. The average value for k is 4.2E-7. [Musa 1998]. w_0 is the initial number of faults in the program and can be calculated using w_0 = N x B where N is the total number of inherent faults and B is the fault to failure conversion rate; that is, proportion of faults that become failures. Proportion of faults not corrected before the software is released. One can assume B = 0.95 i.e 95% of the faults undetected at the time of release and becomes failures after software release. An average of 6 faults per 1000 SLOC can be assumed for w_0 [MUSA 1998].

EXAMPLE 8.1

A software code of 100 SLOC has an average execution rate of 150 lines per second. Predict the failure rate when system test begins.

Using equation (8.3) we have:

$\lambda_0 = f\ k\ w_o$ = (150/100/3) x (4.2 E-7) x (6/1000) = 0.0126E-7 = 1.26E-9 faults per second. The time refers to execution time.

8.4.3 Putnam's model

Trachtenberg (formerly of General Electric) and Gaffney (of then IBM Federal Systems, now Loral) examined the defect histories, by phases of the development process, for many projects of varying size and application type. This model is based on calendar months. Based on their work Putnam assigned the general normalised Rayleigh distribution to describe the observed reliability, where 'k' and 'a' are constants fit from the data and t is time in months:

$$R(t)=k\exp\left(-at^2\right) \tag{8.8}$$

The corresponding probability density function, f(t), the derivative of 1-R(t) with respect to t, is of the general form:

$$f(t)=2akt\ \exp\left(-at^2\right) \tag{8.9}$$

Putnam further developed an ordinal (i.e not equally spaced in real time) scale to represent the development process milestone; (Table 8.1) of special interest is Milestone 7, which is denoted by t_d, corresponding to the end of the development phases and the beginning of full operational capability; this point was defined as occurring at the 95th percentile (i.e. 95% of the defects have been detected at this point in the software development). Using t_d as the reference basis, he then developed the expressions for the model constants, *a* and *k*, in terms of N and t_d. The final equation to predict the number of defects per month as a function of the schedule month and the total number of inherent defects, N is given by:

$$f(t)=\left(\frac{6N}{t_d^{\ 2}}\right)t\exp\left(-\frac{3t^2}{t_d^{\ 2}}\right) \tag{8.10}$$

For example, suppose a software is being developed and the plan is that it will be fully operational (milestone 7) in 10 calendar months resulting in t_d^2 to be 100 (10^2). The defects per month during development are calculated using the equation (8.10):

$$f(t)=0.06\times N\times t\times\exp\left(-0.03t^2\right)$$

Table 8-1. Putnam's time axis milestones

Milestone No.	Milestone
0	Feasibility Study
1	Preliminary Design review, function design complete
2	Critical design review, detailed design complete
3	First code complete
4	Start of systems test
5	Start of users systems test
6	Initial operational capability: installation
7	Full operational capability: reliability about 95% in routine usage
8	99% reliability achieved by stress testing
9	99.9% reliability, assumed debugged

Table 8-2. Expected proportion of the total number of defects

t	f(t)	F(t)	Milestone No
1	0.058	0.058	
2	0.106	0.165	1
3	0.137	0.302	
4	0.149	0.451	2
5	0.142	0.592	
6	0.122	0.715	3
7	0.097	0.811	4
8	0.070	0.881	5
9	0.048	0.929	6
10	0.030	0.959	7
11	0.017	0.976	
12	0.010	0.986	
13	0.005	0.991	8
14	0.002	0.993	
15	0.001	0.994	

Using the expression for f(t) above we can calculate the expected proportion of the total number of defects to be observed in month t. and hence the cumulative proportion F(t). Table 8.2 lists the calculation. The milestone number, based on the planned development schedule is also shown for comparison. Milestone 7, corresponding to the 95th percentile is indeed in month 10, Milestone 8, at the 99th percentile is expected to occur in scheduled month 13, and Milestone 9, at 0.999 is not expected to be reached by the end of scheduled month 15.

This method is useful in predicting the expected number of faults at various points in the development process as compared to Musa's model that provides the prediction only when system testing begins (i.e. at Milestone 4). [RAC-Review 2001]

Another corollary to this model is that the mean time to the next defect MTTD, is given by $1/f(t)$. This is only meaningful after Milestone 4 (since prior to that point the system would not have been developed, so defects cannot be detected). As the development increases (i.e. t increases) the MTTD increases, since defects are being eliminated.

8.5 SOFTWARE RELIABILITY ESTIMATION MODEL

Estimation models use data from the current software development effort. Usually made later in the lifecycle when some data have been collected (not typically used in concept or development phases). These models estimate reliability at either present or at some future time.

Software reliability is estimated during the software testing phase. There are several models available that are concerned with the cumulative number of errors detected by testing (or the time-interval between software failures) and the time span of the software testing and these are called the **software reliability growth models.** These models enable us to estimate software reliability measures such as the mean initial error content, the mean time-interval between failures, the mean number of remaining errors at an arbitrary testing time point, and the software reliability function.

Some of the models which are used for measuring the reliability growth of software are:

- Exponential models
- General Exponential Model
- Goel-Okumoto Model
- Jelinski-Moranda Model

The fault count/fault rate models are the most common type of estimation models. These models are based on the assumptions made on how faults are detected and are corrected.

8.5.1 Exponential model

In general, exponential models assume that the software is in an operational state and that all faults are independent of each other. The time to failure t, of an individual fault follows the exponential distribution:

$$f(t)=\lambda \exp(-\lambda t) \tag{8.11}$$

The reliability R(t) is given by

$$R(t)=\exp(-\lambda t) \tag{8.12}$$

and the Mean time to failure, MTTF, is given by

$$MTTF=\frac{1}{\lambda} \tag{8.13}$$

Let N represent the total number of defects, n the total number of defects detected to date, and c represent the number of defects corrected to date. Then the number of defects yet to be detected equals to (N-n) and the remaining number to be corrected is given by (N-c).

This model is used to estimate both the fault count, n_f, and the fault rate λ_f, at some future time t_f, based on the current (present) value of the fault count, n_p, fault rate, λ_p and time t_p. Typically the count is expressed as the additional count relative to the current value, denoted by Δn. That is, if n_p is the current count and if n_f is the future count, $n_p + \Delta n = n_f$, or $\Delta n = n_f - n_p$. Similarly $\Delta t = t_f - t_p$.

Exponential model is known for its simplicity and its parallelism to the hardware reliability framework. The limitation with this model is that it cannot be used early in the software development since the product must be operational before the model can be used. Hence it cannot be used for early reliability assessment. It also assumes that the number of defects is independent of time and the number already detected and corrected.

8.5.2 Generalized exponential model

In the generalized exponential model, it is assumed that all faults are equal in severity and probability of detection; and that each fault is immediately corrected upon detection. The failure rate λ, is assumed to be directly proportional to the number of faults remaining in the software. That is λ is a function of the number of corrected faults c:

$$\lambda = k(N-c) \tag{8.14}$$

where k is a constant of proportionality. In actual application k can be determined from the slope of the plot of the observed fault rate versus number of faults corrected. The projection of the number of faults to be detected to reach a final failure rate λ_f is given by

$$\Delta n = \left(\frac{1}{k}\right)\left(\frac{\lambda_p}{\lambda_f}\right) \tag{8.15}$$

Where k is the same proportionality constant used above and λ_p is the present failure rate. The projection of the time necessary to reach a projected fault rate is given by:

$$\Delta t = \left(\frac{1}{k}\right)\ln\left(\frac{\lambda_p}{\lambda_f}\right) \tag{8.16}$$

The major limitation of this approach is that the defects be detected and corrected immediately.

8.5.3 Goel-Okumoto Model

Goel–Okumoto software reliability model assumes that the cumulative number of faults detected at time t follows a non-homogeneous Poisson process. The failure intensity is given by:

$$\lambda(t) = \alpha\beta\exp(-\beta t) \tag{8.17}$$

Where α and β are parameters to be determined iteratively from the following::

$$\frac{n}{\alpha} = 1 - \exp(-\beta t) \qquad \textit{and} \qquad \frac{n}{\beta} = \alpha t\exp(-\beta t) + \sum_{i=1}^{n} t_i \tag{8.18}$$

Here α is a number approximately equal to N and β is a number approximately equal to k. N and k are used as starting points for solving the two equations in (8.18) simultaneously. N is the number of inherent faults, and k is the constant of proportionality determined from (8.14)

This model assumes that faults can cause other faults and that they may not be removed immediately and hence an iterative solution is sought. The advantage of this model is that it can be used earlier than other exponential models while its major disadvantage is that it is very sensitive to deviations from the assumptions. The mean value function of Goel-Okumoto Model is given by:

$$m(t)=\alpha[1-\exp(-\beta t)] \tag{8.19}$$

m(t) gives the expected number of failures during t hours of execution.

8.5.4 Jelinski-Moranda Model

One of the earliest models proposed, which is still being applied today, is the model developed by Jenlinski and Moranda, while working on some Navy projects for McDonnell Douglas. In Jelinski Moranda model it is assumed that the failure intensity of the software is proportional to the current fault content. The following assumptions are used:

- The number of faults in the initial software is unknown, but constant.
- The fault that causes a failure is removed instantaneously after failure and no new fault is introduced.
- The failure intensity remains constant throughout the interval between failure occurrences.
- Times between software failures are independent and exponentially distributed.

If N denotes the number of faults present in the initial software, then the failure intensity between the interval $(i-1)^{st}$ and the ith failure is given by

$$\lambda_i = \phi[N - (i-1)], \qquad i = 1,2,\cdots,N \tag{8.20}$$

In equation (8.20) φ is a constant of proportionality denoting the failure intensity contributed by each fault. The distribution of time between failures is given by:

$$P[t_i \le t] = 1 - \exp[-\phi(N - i + 1)t] \tag{8.21}$$

The parameters N and φ can be estimated using maximum likelihood estimators (M Xie, 1991).

8.6 SOFTWARE TESTING

Previous sections have given various models and metrics for prediction of the software reliability. The reliability is entirely dependent on the bugs removed from the software. Only rigorous testing can ensure software reliability. To increase the reliability, it is important to reduce human mistakes by fault-avoidance, e.g. by improved development strategy. In practice due to human unreliability, some mistakes are always made, and by system testing and debugging, we may be able to detect and correct some of these software faults usually caused by human errors. Program proving, especially for safety-critical systems, is not generally possible and program testing has proved to be the only alternative to detect the software faults. There are various techniques for fault-removal purposes. Methods such as module testing, functional testing etc., have been proven to be helpful to detect those software faults that still remain in the software.

If the main program fails during execution, this will be due to one of two possible causes: there are one or more errors in the data or there are one or more errors in the program. Given that the program has been well tested and has been in use for some considerable time, it is more likely that the error is in the data. However, the latter cannot be ruled out completely.

With many computer programming languages, the source code can be compiled in a number of different ways. For fast operation (minimum run times) it is normal to use the (time) optimization option. Unfortunately this generally gives very little diagnostic information following a failure. Sometimes this can be sufficient, particularly if the subroutines are short. But, it would be a lot more helpful if one can find on which line it failed and what values the variables had at that time. To do this, it may be necessary to use a compiler option that gives full checking; however, as mentioned earlier, this considerably increases the execution time of the program.

One possibility is to recompile the program and run the data through it again. A technique that was found to be particularly useful in a fairly large, complex, Monte Carlo simulation was to include a large number of strategically placed write statements which gave the location (in the program), the (simulation) clock time and the values of the 'key' variables. These write statements are by-passed during normal execution but can be activated at certain (simulation clock) times by the user as part of the standard data input.

Another option that is available with Simscript II.5™ is to compile the program with the diagnostics facility enabled. This allows one to selectively 'step through' parts of the program when it is executing giving you the path it has taken and the values of the variables as they are changed within the program. This can be extremely time-consuming process particularly if the failure occurs quite late in to the simulated time. However, it is extremely powerful if all else fails and is particularly useful during the development stages of the model.

A particular problem with simulations which is probably unique to this type of program is that the order in which different parts of the code (typically events or processes) are executed may depend on the sequence of pseudo-random numbers generated by the program and in certain cases, the number of events in what is generally referred to as the "time set" – i.e. the number of events waiting to be executed. This occurs typically while using "heaps" i.e. when the events in the time set are held as a binary tree. It may be possible to avoid this problem by using some form of priority system such that each type of event is given a priority relative to the others. In an early FORTRAN simulation model which used this method, if it suffers an execution error at simulated time T, the normal practice was to set the print switches to come on at some time T – ΔT, say a few days before the error occurs. This, it was hoped, would give the sequence of events, with relevant variable values, leading up to the point of failure. Unfortunately, this new added event (to activate the print switches), could cause a change in the structure of the binary tree resulting in a different path being taken through the code and either causing the error to occur at a different time, or as in many cases, avoiding the error all together. It is perhaps worth pointing out that a novel feature of this particular program was that under certain (ill-defined) circumstances the data storage area would overflow into the memory area used to store the executable code with often quite bizarre, not to say unpredictable, results. Fortunately when the model was rewritten in SIMSCRIPT II.5 this feature was not duplicated.

Experience with this type of modeling makes one aware of the importance of having good data input routines that check the data being input. Ideally, these should restrict the user to as small a set of permitted values as possible. Thus, for example, if the program is expecting to read a probability, it should make sure the user cannot enter a value less than 0 or greater than 1. If it requires a vector of values which divide the options into a set of mutually exclusive and exhaustive alternatives then the model should check the sum of these values is either 1 or 100 (depending on whether it is expecting them to be probabilities or percentages). In other words, lot of execution errors can be avoided if the users are forced to operate within the programs operating envelop or boundaries, just as pilots are usually discouraged from over revving the engine or pulling more than a certain G-force.

8.7 SOFTWARE RISK ANALYSIS USING FMEA

FMEA can be used to assess the quality risks in developing software and also as an aid for functional testing. For software, FMEA would serve as a top-down process for understanding and prioritizing possible failure modes in system function and features. Table 8.3 shows a sample FMEA chart. The description of the columns in the table is as follows.

The first three columns describe the system function whose failure modes are analyzed with the effects of each mode of failure. The criticality of the effect is answered with a YES or NO. Is the product feature or function completely unusable if the particular failure mode occurs? The potential causes of the failure mode are also listed. This column lists the possible factors that might trigger the failure, for example, operating system error, user error or normal use.

To calculate the Risk Priority Number (RPN) numerical values are entered in the scale of 1 (worst) to 5 (least dangerous) in Severity, Priority and Detection columns.

Severity: This column denotes the absolute severity of the failure mode in question, regardless of likelihood.

1. Loss of data or safety Issue
2. Loss of functionality with no workaround
3. Loss of functionality with a workaround
4. Partial loss of functionality
5. Cosmetic or trivial

Priority: This rates how badly the failure mode will affect the company's ability to market the product during its expected life cycle, assuming that the failure mode remains in the software product on release.

1. Implies most dangerous
2. .
3. .
4. .
5. Least dangerous

Detection: The number in this column represents the likelihood of detection by the methods listed in the detection method column. In other words if we are confident that current procedures will catch this failure mode, the column should indicate a low risk. Therefore,

1. Most likely to escape
2. .
3. .
4. .
5. Least likely to escape current detection

Risk Priority Number is the product of the three numbers entered in the severity, priority and detection columns. Since the numbers entered in each column ranges from 1 to 5 the RPN will vary from 1 to 125 indicating 1 (high risk) and 125(least risk).

The recommended action lists simple action for increasing the risk priority number (making it less severe). Based on the RPN the test team can create test cases that influence the detection figure. The rest of the column describes the person responsible for each recommended action, certain references for more information about the RPN. The RPN calculation can be repeated after the recommended actions are implemented to indicate the improvement or reduction in the risk.

Although RPN technique is widely used in the hardware and software to prioritize the critical failures, one should use them carefully. The RPN is the product of three factors severity, priority and detection. The larger value of one of the factors will increase the RPN indicating that the problem is less severe. Thus while prioritizing the risks one should pay attention to the individual parameter especially, to severity.

EXAMPLE 8.2

An example of FMEA for a specific function in aircraft cockpit display is given in Table 8.3. The cockpit display has several major functions such as multifunction display, processing and Head Up Display. Page handler is a specific function, which aids in displaying information on several attributes such as fuel requirement, engine status, and electrical parameters on several pages after processing the data received from various sensors. Table describes certain modes of failures, their severities and hence the risk priority number. Based on the risk priority numbers one can prioritize the test cases to be written for various failure modes of the function considered.

FMEA involves considerable effort in terms of time and cost. Thus one may probably want to limit the analysis to highly used and/or most critical functions. The FMEA can be applied at the functional level and the microscopic level analysis with the component (in this case variables) can be carried out for potential failures leading to severity I. Failure prevention can be implemented for those items. Conducting FMEA at the module level can help you identify which modules are most critical to the proper system operations.

8.8 FAULT TOLERANT SOFTWARE

Software is said to be fault tolerant if it operates satisfactorily in the presence of faults (defects that would cause failures that are unsatisfactory to operation, if not counteracted). The malfunction of a fault-tolerant feature, which is a requirement that a system defend against certain faults, is a failure. Two common fault-tolerant techniques applied to software are multi-version programming (N-version programming) and recovery blocks. It is essential that the code developed for fault tolerant purpose is error free. Both fault tolerant architectures, N-version programming and recovery block achieve fault tolerance by using redundancies within the software.

8.8.1 Recovery blocks

A recovery block scheme requires n independent versions of a program and a testing segment, where different versions of a program are executed sequentially. Upon invocation, the recovery block first executes the first version of the recovery block. The output from this version is submitted to a testing segment, which may either accept or reject the output. If an output is rejected (this indicates the failure of a version), the testing segment activates the succeeding version for execution, and this process continues until the all the versions are exhausted. Testing segment is a software and its function is to ensure that the operation performed by a version is correct. If the output of a version is incorrect then the testing segment recovers the initial state before activating the next version (if one is available). The initial state refers to the values of global variables just before entering the recovery block. Since the testing segment is again a software, it can also fail. The reliability of a recovery block can be derived as follows:

Let:

N = the number of versions in a recovery block

P_i = Failure probability of version i.

T_1 = Probability that the testing segment cannot perform successful recovery of the initial state upon failure of a version.

T_2 = Probability that the testing segment rejects correct output.

T_3 = Probability that the testing segment accepts incorrect output.

All the versions of the recovery block are assumed to be independent. To achieve statistical independence, the different versions of the recovery block are developed by independent teams using different design and logic. Three possible errors that can result in failure of a recovery block are:

1. A version produces a correct output, but the testing segment rejects it.
2. A version produces an incorrect output, but the testing segment accepts it.
3. The testing segment cannot perform successful recovery of the initial state upon failure of a version.

We define the following two events:

X_i is the event that either version *i* produces an incorrect output and the testing segment rejects it or the version produces a correct output and the testing segment rejects it; in either case the testing segment is able to perform successful recovery of the initial state.

Y_i is the event that version *i* produces a correct output and the testing segment accepts the correct output.

Probabilities of the events X_i and Y_i are given by:

$$P(X_i) = (1 - T_1)[P_i(1 - T_3) + (1 - P_i)T_2]$$

$$P(Y_i) = (1 - P_i) \times (1 - T_2)$$

Using the above notation, the reliability of a recovery block with 'n' versions, R_n, is given by (Berman and Dinesh Kumar, 1999):

$$R_n = P(Y_1) + \sum_{i=2}^{n} \left(\prod_{k=2}^{i-1} P(X_k) \right) P(Y_i) \tag{8.22}$$

For n > 1, the reliability of a recovery block can be calculated using the following recursive equation.

$$R_n = R_{n-1} + \left[\prod_{k=1}^{n-1} P(X_k) \right] P(Y_n) \qquad (8.23)$$

8.8.2 N-version programming

In an N-version programming, N independently coded versions of a program are executed concurrently and the results are compared using a majority voting mechanism. The concept of N-version programming was first introduced in 1977 by Avizienis and Chen (1977) and is defined as:

"The independent generation of $N \geq 2$ functionally equivalent programs from the same initial specification. The N programs possess all the necessary attributes for concurrent execution during which comparison vectors (c-vectors) are generated by programs at certain points. The program state variables that are to be included in each c-vector and the cross check points at which c-vectors are to be generated are specified along with the initial specifications".

N-version programming is a popular fault tolerant architecture which is used in many safety critical systems such as space shuttles and fighter aircrafts. A specific limitation of N-version programming is that it requires N computers that are hardware independent and able to communicate very efficiently to compare the outputs. Assume that there are 3 versions in a N-version programming and at least two versions should produce same output for the success of a N-version programming. Then the reliability of a N-version programming is given by:

$$R_n = \left[(1-P_1)(1-P_2)P_3 + (1-P_1)(1-P_3)P_2 + (1-P_2)(1-P_3)P_1\right] \times PR \qquad (8.24)$$

Where P_i is the failure probability of version i and PR is the probability that a recurring output is correct.

8.9 CASE STUDY: ARIANE 5

On 4 June 1996, Ariane 5, a giant rocket developed by European Space Agency, exploded within 39 seconds of its launch with four expensive and uninsured scientific satellites. It took the European Space Agency around 10 years and $7 billion to make Ariane 5. The failure was caused by the complete loss of guidance and attitude information. The loss of information and guidance was due to specification and design errors in the software of the inertial reference system (IRS). One of the main recommendations of the commission set up to investigate the crash was that, 'no software function should run during flight unless it is needed'. In a way, it also talks about the confidence people have on software. In fact, the same commission went on to say that it is impossible to make 100% reliable software.

The main function of Inertial Reference System (IRS) is to measure the attitude of the launcher and its movements in space. The embedded computer within the IRS calculates angles and velocities on the basis of information from an inertial platform with laser gyros and accelerometers. The data from the IRS are transmitted to the on-board computer, which executes the flight program. To improve the reliability of IRS, there is a parallel redundancy. That is two IRS operate in parallel with identical hardware and software. One IRS is used in active mode and other is in hot standby mode. If the on-board computer detects that the active IRS has failed, it immediately switches to the other one, provided that the standby unit is functioning properly. During the flight the active IRS failed due to the software error, this was caused during execution of a data conversion from 64 bit floating point to 16 bit signed integer value. The floating point value which was converted had a value greater than what could be represented by a 16 bit signed integer. This resulted in an operand error and unfortunately the data conversion instructions were not protected. When the guidance system shut down, it passed control to the redundant IRS. However, the hot standby IRS, which was there to provide backup, had failed in the identical manner a few milliseconds before.

During any product development, several decisions are made. During the development of the software, not all conversions are protected. The software is written in Ada, which is believed to be more secure compared to other programming languages such as Fortran or C++. Ironically, If European Space Agency had chosen, C or C++ the data conversion would have overflowed silently, and crash may not have occurred.

Reliability of software involves calculating reliability of all the processes used in developing that software. Starting from requirement stage till the final acceptance, all the processes used finally determine the reliability of the software. In case of Ariane 5, the part of the software which caused the failure in the IRS is used in Ariane 4, to enable a rapid realignment of the system in case of a late hold in the countdown. This realignment function which didn't serve any purpose on Ariane 5, was nevertheless used to retain the commonality.

Although, there was redundant IRS system, the software used in both on-board computers was same. Unlike in hardware, duplicating software will never improve the reliability. To achieve fault tolerance in software, it is essential to make sure that the different versions of the software are independently developed using different logic and design. A true N-version programming architecture could have saved Ariane 5 from crashing.

Table 8-3. Failure mode and effects analysis - risk analysis form

System function or feature	Potential failure modes	Potential effects of failure	Criticality (Y/N)	Severity (1 to 5)	Potential Cause	Priority	Detection Methods	Detection (1 to 5)	RPN	Recom-mended action	Who? When?	Reference
System name: Cockpit display System Responsibility: Person Responsibility: Involvement of others:				Supplier involvement: Model/Product: Target Release Date: Prepared by:					FMEA Date: FMEA Rev. Date			
Page Handler To aid the pilot with different pages of informati-on such as fuel page, electrical page, engine page	Date flow time not OK (20 milliseco-nds not met)	Data Distortion	Y	1	Slow Process	2	Test with tolerance wiring/right	4	8	Toleran-ce provided		
	Wrong data received. Error in realizing the informati on recticles	Wrong informati on on fuel, electrical parameter , engine rpm etc	Y	1	Validity checks not OK	1	A bit for every page introduced for check	4	4	Introduc e more validity checks		
	System hanging	No output	Y	3		2	Watch dog timer	4	24	System reboot automati cally		

Chapter 9

AVAILABILITY AND SIX SIGMA

Informed decision making comes from a long tradition of guessing and then blaming others for inadequate results.
Scott Adams

9.1 INTRODUCTION

Availability is a measure of system performance and measures the combined effect of reliability, maintenance and logistic support on the operational effectiveness of the system. A system which is in a state of failure is not beneficial to its user; in fact, it is probably costing the user money. If an aircraft breaks down, it cannot be used until it has been declared airworthy. This is likely to cause inconvenience to the customers who may then decide to switch to an alternative airline in future. It may disrupt the timetables and cause problems for several days. The availability can be easily converted in to a Sigma level quality. Once again, we define the defects as the additional number of systems that are not available due to not meeting the target availability. For example, if the target availability for a system is 0.95 and the observed availability is 0.90, then one at the fleet level can expect 5% additional systems unavailable due to not meeting the target availability. That is, out of 100 units, 5 additional units will not be available. The DPU (defect per unit) equivalent of this is 0.05. The corresponding Sigma level quality is 3.14.

Availability is probably the most important measure of system performance for many users. For example, most large airlines have a very high utilization rate with the only down time being to do a transit check, unload, clean the cabin, refuel, restock with the next flight's food and other items, and reload with the next set of passengers and baggage. The whole operation generally takes about an hour. Any delay may cause it to miss its take off slot and more significantly its landing slot, since an aircraft cannot take-off until it has been cleared to land, even though this may be 12 hours later. Many airports close during the night to avoid unacceptable levels of noise pollution. If the particular flight was due to land just before the airport closes, missing its slot could mean a delay of several hours.

An operator of a system would like to make sure that the system will be in a state of functioning (*SoFu*) when it is required. Designers and manufacturers know that they are unlikely to remain in business for very long if their systems do not satisfy the customers' requirements in terms of operational effectiveness. Many forms of availability are used to measure the effectiveness of the system. Inherent availability, operational availability and achieved availability are some of the measures used to quantify whether an item is in an operable state when required. Availability is defined as:

The probability that an item is in the state of functioning at a given point in time (point availability) or over a stated period of time (interval availability) when operated, maintained and supported in a prescribed manner.

In this chapter, we look at few important availability measures such as point availability, interval availability, steady state inherent availability, operational availability and achieved availability and how one can calculate the Sigma quality level based on the target and observed availability values.

9.2 POINT AVAILABILITY

Point availability is defined as the probability that the system is in the state of functioning (*SoFu*) at the given instant of time t. We use the notation $A(t)$ to represent the point availability. Availability expressions for systems can be obtained by using stochastic processes. Depending on the time to failure and time to repair distributions, one can use Markov chain, renewal process, regenerative process, semi-Markov process and semi-regenerative process models to derive the expression for point availability.

9.2.1 Markov model for point availability

Consider an item with constant failure rate λ and constant repair rate μ. At any instant of time, the item is either in the state of functioning (say, state 1) or in the state of failure (say, state 2). As both failure and repair rates are constant (and thus follow exponential distribution), we can use Markov chain to model the system to derive the availability.

Let $p_{ij}(h)$ denotes the transition probability from state i to state j during the interval 'h'$(i,j = 1, 2)$. Define, $P_i(t+h)$, as the probability that the system would be in state i at time $t+h$, for $i = 1, 2$. The expression for $P_1(t+h)$ can be derived using the following logic:

1. The system was in state 1 at time t and continues to remain in state 1 throughout the interval h.
2. The system was in state 2 at time t and it transits to state 1 during the interval h.

The corresponding expression can be written as:

$$P_1(t+h) = P_1(t) \times p_{11}(h) + P_2(t) \times p_{21}(h) \quad (9.1)$$

Using similar logic, the expression for $P_2(t+h)$ can be written as:

$$P_2(t+h) = P_1(t) \times p_{12}(h) + P_2(t) \times p_{22}(h) \quad (9.2)$$

$p_{11}(h)$ is the probability of remaining in state 1 during the interval h. The probability $p_{11}(h)$ is given by

$$p_{11}(h) = \exp(-\lambda h) \approx 1 - \lambda h \text{ for } \lambda h << 1$$

$p_{21}(h)$ is the probability of entering state 1 from state 2 during the interval h. The corresponding expression is given by

$$p_{21}(h) = 1 - \exp(-\mu h) \approx \mu h \text{ for } h\mu << 1$$

$p_{12}(h)$ is the probability of entering state 2 from state 1 during the interval h. The probability $p_{12}(h)$ is given by

$$p_{12}(h) = 1 - \exp(-\lambda h) \approx \lambda h \text{ for } h\lambda << 1$$

$p_{22}(h)$ is the probability of remaining in state 2 during the interval h. The probability $p_{22}(h)$ is given by:

$$p_{22}(h) = \exp(-\mu h) \approx 1 - \mu h \text{ for } h\mu << 1$$

Substituting the values of $p_{ij}(h)$ in equation (9.1) and (9.2), we get

$$P_1(t+h) = P_1(t) \times (1 - \lambda h) + P_2(t) \times \mu h$$

$$P_2(t+h) = P_1(t)\times(\lambda h) + P_2(t)\times(1-\mu h)$$

By rearranging the terms and setting $h \to 0$, we have

$$\underset{h\to 0}{Lt}\frac{P_1(t+h)-P_1(t)}{h} = \frac{dP_1(t)}{dt} = -\lambda P_1(t) + \mu P_2(t)$$

$$\underset{h\to 0}{Lt}\frac{P_2(t+h)-P_2(t)}{h} = \frac{dP_2(t)}{dt} = \lambda P_1(t) - \mu P_2(t)$$

On solving the above two differential equations, we get

$$P_1(t) = \frac{\mu}{\lambda+\mu} + \frac{\lambda}{\lambda+\mu}\times\exp(-(\lambda+\mu)t)$$

$P_1(t)$ is nothing but the availability of the item at time t, that is the probability that the item will be in state of functioning at time t. Thus, the point availability $A(t)$ is given by:

$$A(t) = \frac{\mu}{\lambda+\mu} + \frac{\lambda}{\lambda+\mu}\times\exp(-(\lambda+\mu)t) \tag{9.3}$$

Substituting $\lambda = 1/MTTF$ and $\mu = 1/MTTR$ in the above equation, we get

$$A(t) = \frac{MTTF}{MTTF+MTTR} + \frac{MTTR}{MTTF+MTTR}\times\exp(-(\frac{1}{MTTF}+\frac{1}{MTTR})t) \tag{9.4}$$

When the time to failure and time to repair are not exponential, we can use a *regenerative process* to derive the availability expression. If $f(t)$ and $g(t)$ represent the time-to-failure and time-to-repair distributions respectively, then the point availability $A(t)$ can be written as (Birolini, 1997):

$$A(t) = 1 - F(t) + \int_0^t \sum_{n=1}^{\infty} [f(x)*g(x)]^n [1-F(t-x)]dx$$

where $[f(x)*g(x)]^n$ is the n-fold convolution of $f(x)*g(x)$. The summation $\sum_{n=1}^{\infty}[f(x)*g(x)]^n$ gives the renewal points $f(x)*g(x)$, $f(x)*g(x)*f(x)*g(x)$, ...

lies in $[x, x+dx]$, and $1 - F(t-x)$ is the probability that no failures occur in the remaining interval $[x, t]$.

9.2.2 Average availability

Average availability, $AA(t)$, is defined as the expected fractional duration of an interval $(0, t]$ that the system is in state of functioning. Thus,

$$AA(t) = \frac{1}{t}\int_0^t A(x)dx \tag{9.5}$$

Where $A(x)$ is the point availability of the item as defined in equation (9.3) and (9.4). For an item with constant failure rate λ and constant repair rate μ, the average availability is given by:

$$AA(t) = \frac{\mu}{\lambda+\mu} + \frac{\lambda}{(\lambda+\mu)^2 t}[1-\exp(-(\lambda+\mu)t)] \tag{9.6}$$

9.2.3 Inherent availability

Inherent availability (or steady-state availability), A_i, is defined as the steady state probability (that is, $t \to \infty$) that an item will be in a state of functioning, assuming that this probability depends only on the time-to-failure and time to repair distributions. It is assumed that any support resources that are required are available without any restriction. Thus, the inherent availability is given by:

$$A_i = \underset{t\to\infty}{Lt}\ A(t) = \frac{MTTF}{MTTF+MTTR} \tag{9.7}$$

The above result is valid for any time to failure function $F(t)$ and any time to repair distribution $G(t)$ (*Birolini*, 1997). Also, in the case of constant failure rate λ and constant repair rate μ, the following inequality is true.

$$\left|A(t) - A_i\right| \le \exp(-t/MTTR) \tag{9.8}$$

9.2.4 Availability, DPMO and Sigma level

As already mentioned, Inherent availability is an important measure of system operational effectiveness. During the initial stages of product development, it is necessary to set a target for inherent availability, and the design activities should make sure that they try to achieve the target inherent availability. Any deviation from the target, especially towards the lower side, is likely to increase the total cost of ownership and restriction of operation to the user. The defect per unit, DPU, in this case is defined using the following equation:

$$DPU = A_i - A_{oi} \qquad \text{for } A_{oi} < A_i \tag{9.9}$$

Where A_i is the target inherent availability (as claimed by the original equipment manufacturer) and A_{oi} is the observed inherent availability.

The corresponding DPMO is given by:

$$DPMO = (A_i - A_{oi}) \times 10^6, \qquad \text{for } A_{oi} < A_i \tag{9.10}$$

The corresponding process sigma level is given by:

$$\text{Sigma level} = 0.8406 + \sqrt{29.37 - 2.221 \times \ln\left((A_i - A_{oi}) \times 10^6\right)} \tag{9.11}$$

The corresponding Excel® is given by:

NORMINV(1 – (A_i – A_{oi}), 1.5, 1)

Note: Inherent availability is a measure of reliability and maintainability. The original equipment manufacturer should make sure that the reliability and maintainability targets are met during the product development to achieve the target inherent availability. If the original equipment manufacturer fails to achieve the target inherent availability, then there will be more unavailable products compared to the case where the target availability is met. The excess number of unavailable systems are treated as defects in our calculation of process sigma.

EXAMPLE 9.1

Time to failure distribution of a digital engine control unit (DECU) follows an exponential distribution with mean time between failures 1200 hours and the repair time also follows an exponential distribution with mean time to repair 400 hours.

1. Plot the point availability of the DECU.
2. Find the average availability of the DECU during the first 5000 hours.
3. Find the inherent availability.

The point availability of the DECU is calculated using the equation (9.4). Figure 9.1 depicts the point availability of the system.

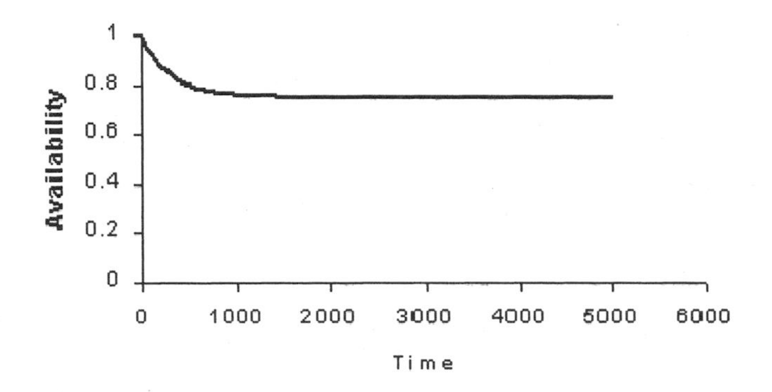

Figure 9-1. Point availability of DECU

The average availability of the system during 5000 hours of operation is given by:

$$AA(t) = \frac{\mu}{\lambda + \mu} + \frac{\lambda}{(\lambda + \mu)^2 t}[1 - \exp(-(\lambda + \mu)t)]$$

Substituting the values of λ (= 1/1200) and μ (=1/400), we get the value of the average availability during 5000 hours as 0.7649.

The inherent availability is given by

$$A_i = \frac{MTTF}{MTTF + MTTR} = \frac{1200}{1200 + 400} = 0.75$$

Thus, the steady state availability of the system is 0.75 or 75%.

EXAMPLE 9.2

Assume that the target inherent reliability (or availability claimed by the original equipment manufacturer) for the digital engine control unit mentioned in example 9.1 is 0.82. Calculate the DPMO, where the defect is the additional number of units that are likely to be unavailable due to not meeting the target availability and the corresponding Sigma level.

Solution:

The defect per unit (DPU) = 0.82 – 0.75 = 0.07

The corresponding DPMO = 70,000

The process sigma level is given by (equation 9.11):

$$\text{Sigma level} = 0.8406 + \sqrt{29.37 - 2.221 \times \ln(70000)} = 2.98$$

9.2.5 System Availability

Availability of a system with series reliability block diagram with n items is given by

$$A_s(t) = \prod_{k=1}^{n} A_i(t) \tag{9.12}$$

where $A_i(t)$ is the point availability of ith item. The inherent availability of the system is given by

$$A_{i,s} = \prod_{k=1}^{n} \frac{MTTF_i}{MTTF_i + MTTR_i} \tag{9.13}$$

For a series system with all the elements having constant failure and repair rates, the system inherent availability

$$A_{i,s} = \frac{MTTF_s}{MTTF_s + MTTR_s} \tag{9.14}$$

$MTTF_s$ and $MTTR_s$ are system mean time to failure and system mean time to repair respectively. Let λ_i and μ_i represent the failure rate and repair rate of item i respectively. $MTTF_s$ and $MTTR_s$ are given by

$$MTTF_s = \frac{1}{\sum_{i=1}^{n} \lambda_i}$$

$$MTTR_s = \sum_{i=1}^{n} \frac{\lambda_i MTTR_i}{\lambda_s}, \text{ where } \lambda_s = \sum_{i=1}^{n} \lambda_i$$

Availability of a parallel system with n items is given by

$$A_s(t) = 1 - \prod_{i=1}^{n} [1 - A_i(t)] \tag{9.15}$$

EXAMPLE 9.3

A series system consists of four items. The time to failure and the time to repair distributions of the different items are given as given in Tables 9.1 and 9.2. Find the inherent availability of the system.

Table 9-1. Time to failure distribution of different items

Item Number	Distribution	Parameters
Item 1	Weibull	η = 2200 hours β = 3.7
Item 2	Exponential	λ = 0.0008 per hour
Item 3	Weibull	η = 1800 hours β = 2.7
Item 4	Normal	μ = 800 hours σ = 180 hours

Table 9-2. Time to repair distribution of different items

Item number	Distribution	Parameters
Item 1	Lognormal	$\mu_l = 3.25$ and $\sigma_l = 1.25$
Item 2	Normal	$\mu = 48$ hours $\sigma = 12$ hours
Item 3	Lognormal	$\mu_l = 3.5$ and $\sigma_l = 0.75$
Item 4	Normal	$\mu = 72$ hours $\sigma = 24$ hours

First we calculate $MTTF_i$ and $MTTR_i$ for different items:

$$MTTF_1 = \eta \times \Gamma(1+\frac{1}{\beta}) = 2200 \times \Gamma(1+\frac{1}{3.7}) = 2200 \times 0.902 = 1984.4$$

$$MTTF_2 = 1/\lambda = 1/0.0008 = 1250 \text{ , } MTTF_3 = 1600.2, \quad MTTF_4 = 800$$

$$MTTR_1 = \exp(\mu_l + \sigma_l^2/2) = 56.33 \text{ hours}, \quad MTTR_2 = 48 \text{ hours}$$

$$MTTR_3 = \exp(\mu_l + \sigma_l^2/2) = 43.87 \text{ hours}, \ MTTR_4 = 72 \text{ hours}$$

Inherent availability, A_i, for item i can be calculated using the equation (9.7). Substituting the values of $MTTF_i$ and $MTTR_i$ in equation (9.7), we have:

$$A_1 = 0.9723, \quad A_2 = 0.9630, \ A_3 = 0.9733, \ A_4 = 0.9174$$

The system availability is given by:

$$A_s = \prod_{i=1}^{4} A_i = 0.8362$$

EXAMPLE 9.4

In example 9.3, assume that the target system availability is 0.85. Calculate the DPMO and Sigma level for the system. Using equal apportionment technique, calculate the individual availability target to meet the system level availability target of 0.85.

The defect per unit (DPU) = 0.85 – 0.8362 = 0.0138

The corresponding DPMO = 13800

The process sigma level is given by (equation 9.11):

$$\text{Sigma level} = 0.8406 + \sqrt{29.37 - 2.221 \times \ln(13800)} = 3.70$$

Now using the equal apportionment technique, the individual availabilities, A_i, are given by:

$$A_i = (A_{system})^{1/4} = (0.85)^{1/4} = 0.960185$$

9.3 ACHIEVED AVAILABILITY

Achieved availability is the probability that an item will be in a state of functioning (*SoFu*) when used as specified taking into account the scheduled and unscheduled maintenance; any support resources needed are available instantaneously. Achieved availability, A_a, is given by:

$$A_a = \frac{MTBM}{MTBM + AMT} \tag{9.16}$$

MTBM is the mean time between maintenance and AMT is active maintenance time. The mean time between maintenance during the total operational life, T, is given by:

$$MTBM = \frac{T}{M(T) + T/T_{sm}} \tag{9.17}$$

M(T) is the renewal function, that is the expected number of failures during the total life T. T_{sm} is the scheduled maintenance interval (time between scheduled maintenance). The above expression is valid when after each scheduled maintenance, the item is 'as-bad-as-old' and after each corrective maintenance the item is 'as-good-as-new'. The active maintenance time, AMT, is given by:

$$AMT = \frac{M(T) \times MTTR + (T / T_{sm})MSMT}{M(T) + T / T_{sm}} \tag{9.18}$$

MTTR stands for the mean time to repair and MSMT is the mean scheduled maintenance time.

EXAMPLE 9.5

Time to failure distribution of an engine monitoring system follows a normal distribution with mean 4200 hours and standard deviation 420 hours. The engine monitoring system is expected to last 20,000 hours (subject to corrective and preventive maintenance). A scheduled maintenance is carried out after every 2000 hours and takes about 72 hours to complete the task. The time to repair the item follows a lognormal distribution with mean time to repair 120 hours. Find the achieved availability for this system.

Mean time between maintenance, MTBM, is given by

$$MTBM = \frac{T}{M(T) + T / T_{sm}} = \frac{20000}{M(20000) + 20000 / 2000}$$

M(20000) for normal distribution with mean 4200 hours and standard deviation 420 hours is given by

$$M(20000) = \sum_{n=1}^{\infty} \Phi(\frac{20000 - n \times 4200}{\sqrt{n} \times 420}) = 4.1434$$

$$MTBM = \frac{20000}{4.1434 + 10} \approx 1414 \text{ hours}$$

The active maintenance time is given by:

$$AMT = \frac{M(T) \times MTTR + (T/T_{sm})MSMT}{M(T) + T/T_{sm}}$$

$$= \frac{4.1434 \times 120 + 10 \times 72}{4.1434 + 10} \approx 86.06$$

The achieved availability of the system is given by:

$$A_a = \frac{MTBM}{MTBM + AMT} = \frac{1414}{1414 + 86.06} = 0.9426$$

9.4 OPERATIONAL AVAILABILITY

Operational availability is the probability that the system will be in the state of functioning (*SoFu*) when used as specified taking into account maintenance and logistic delay times. Operational availability, A_o, is given by

$$A_o = \frac{MTBM}{MTBM + DT} \tag{9.19}$$

Where, MTBM is the mean time between maintenance (including both scheduled and unscheduled maintenance) and DT is the Down time. The mean time between maintenance during the total operational life, T, is given by:

$$MTBM = \frac{T}{M(T) + T/T_{sm}} \tag{9.20}$$

M(T) is the renewal function, that is the expected number of failures during the total life T. T_{sm} is the scheduled maintenance interval (time between scheduled maintenance). The system down time DT is given by:

$$DT = \frac{M(T) \times MTTRS + (T/T_{sm})MSMT}{M(T) + T/T_{sm}} \tag{9.21}$$

MTTRS stands for the mean time to restore the system and MSMT is the mean scheduled maintenance time. MTTRS is given by

MTTRS = MTTR + *MLDT*

where MLDT is the mean logistic delay time for supply resources. In the absence of any scheduled maintenance the operational availability can be calculated using the following simple formula

$$A_o = \frac{MTBF}{MTBF + MTTR + MLDT} \tag{9.22}$$

EXAMPLE 9.6

Consider a system which is fixed only when it is broken. The mean time between failures of the system is 1789 hours. On average it takes about 156 hours to recover the system and the mean logistic delay time is 96 hours. Calculate the operational availability of the system. If the target operational availability is 0.90, calculate the DPMO and the sigma level.

Solution:

The operational availability is given by:

$$A_o = \frac{MTBF}{MTBF + MTTR + MLDT} = \frac{1789}{1789 + 156 + 96} = 0.8765$$

The defect per unit (DPU) = 0.90 – 0.8765 = 0.023469

The corresponding DPMO = 23469

The process sigma level is given by (equation 9.11):

$$\text{Sigma level} = 0.8406 + \sqrt{29.37 - 2.221 \times \ln(23469)} = 3.4899$$

Note: The sigma level in the case of operational availability is not only due to the process capabilities of the original equipment manufacturer, but also due to the process capability of the user. The mean logistic delay mainly depends on the process used by the user in providing the logistic resources to recover the failed item.

Chapter 10

RELIABILITY AND SIX SIGMA MANAGEMENT

Change before you have to.

Jack Welch

10.1 INTRODUCTION

Reliability is probably the most important design characteristic. All other system characteristics such as maintenance, spare parts, logistic support and the cost of ownership depend heavily on the reliability of the product. Unfortunately, apart from very few industries such as defence, aerospace and nuclear, the knowledge of reliability about their own product is very limited among many industries. In fact, for many the reliability stops with MTBF. Unfortunately, many important product related decisions are made based on this most unreliable reliability measure. Most people are even unaware that only one out of three products would survive beyond its MTBF value (assuming exponential distribution). Failure to understand the true reliability would hurt any manufacturer, especially if there is a strong competition in the aftermarket. Product aftermarket spans the life cycle of the product between the end of the original equipment manufacturer's (OEM) warranty and till its end of life when the product is scrapped. Aftermarket generates 40-50% of a manufacturing companies profit and 25% of its annual revenue. Aftermarket is a very important source of revenue since many manufacturers do not make profit when they actually sell the product; the actual profit comes when they start selling the spare parts. The demand for spare parts heavily depends on the reliability of the product and thus failure to understand the reliability of the product will result in poor prediction being made of the demand of spare parts. Poor prediction about the demand for spare parts can result in loss either due to overstocking inventory or due to lost sales.

Reliability and Six Sigma projects have many things in common. To start with most of the tools used in reliability projects are also used in Six Sigma projects. Increasing product reliability is one of the popular Six Sigma projects across many companies. In this chapter the management aspects of reliability and Six Sigma will be discussed.

10.2 RELIABILITY MANAGEMENT

Managing a reliability programme within a product development scenario should typically start from the conceptual design stage of the product development. Any reliability management program should include the following steps:

1. Definition of quantitative targets for reliability during conceptual design stage. The quantitative target should reflect the operational requirement of the product.
2. Preparation of reliability block diagrams at various indenture levels (hierarchy of engineering systems such as assembly, module, parts etc).
3. Allocation of reliability at various indentures (assemblies, modules and parts)
4. Identification of product weaknesses using FTA and FMECA
5. Reliability improvement through redundancy.
6. Reliability prediction at various indenture levels.
7. Selection of materials for in-house manufacturing and vendors of outsourced assemblies, modules and parts.
8. Reliability demonstration.
9. Reliability growth programs.

In this chapter we discuss items 8 and 9 in the above list. Other issues have been already discussed in various chapters of this book.

10.2.1 Reliability demonstration

Reliability demonstration is a process of demonstrating product reliability using various reliability tests. The tests are designed to verify that the product meets the specified reliability. The reliability demonstration tests should be designed to simulate the operational conditions as close as possible. In the early concept and design stages, the only reliability data available is the target MTBF for the system which may be expressed as x failures per thousand operational hours or, more recently, as a probability of surviving a given period without the need for any corrective or preventative maintenance. There may be some feedback from similar systems currently in operation or recently retired. However, in most cases the prospective buyers and operators will be looking for a significant improvement on past (reliability) performance so this data may be of limited applicability.

The Reliability Demonstration Testing (RDT) is also known as Reliability Qualification Testing (RQT). The latter title is often used to designate testing imposed by a customer. Demonstration of reliability might be required as part of development and production contract, to ensure the requirements have been met. In either case, the purpose is to provide a specified degree of confidence that the desired reliability has been achieved. The reliability demonstration involves the following two steps:

1. Estimation of unknown parameters such as probability of failure, failure, MTTF etc. The estimation can be either a point estimate or an interval estimate of an unknown parameter.
2. Once the estimate of the parameter is known, testing a given hypothesis on the unknown parameter in an acceptance test.

10.2.1.1 Two-Sided Reliability Demonstration Test

The simplest reliability demonstration test is based on binomial distribution and is used to demonstrate the defective probability p. Suppose a customer is planning to procure n number of products from a producer. While accepting the products the customer would normally like to ensure that none of the products fail. But it will be a very stringent condition on the producer. The question arises as to how many failures are acceptable for the customer. This is usually specified using defective probability p_0 at a certain confidence level. The main objective of the two sided reliability demonstration test is to check a null hypothesis H_0: $p < p_0$ against the alternative hypothesis H_1: $p > p_1$. The lot should be accepted with a probability 1- α if the true defective probability p is lower than p_0 and rejected with probability 1 - β if p is greater than p_1. Where p_0 is specified defective probability and p_1 is the maximum acceptable defective probability (Birolini, 1994). Here α denotes the probability of rejecting a null hypothesis (called producers risk or type I error) and β is consumers risk (type II error), that is probability of accepting a null hypothesis although the alternative hypothesis is true. Since the number of failures follow binomial distribution, one can use the following test plan for reliability demonstration (Birolini, 1994).

1. Find the smallest integers c and n for which

$$\sum_{i=0}^{c} \binom{n}{i} p_0{}^i (1-p_0)^{n-1} \geq 1-\alpha \tag{10.1}$$

and

$$\sum_{i=0}^{c} \binom{n}{i} p_1{}^i (1-p_1)^{n-1} \leq \beta \tag{10.2}$$

2. Take a sample of size n and determine the number of defective items k in the sample and reject the null hypothesis H_0 ($p < p_0$) if k is greater than c or accept the null hypothesis if k is less than or equal to c.

Note that the defective probability can be replaced with one minus system reliability at time t, R(t) [that is, we can substitute p = 1-R(t)]. Let $R_0(t)$ be the specified reliability and $R_1(t)$ be the minimum acceptable reliability. The number of failures, N_f, by time t is a random variable. Assume that there are n units in the sample. Then from the argument stated in the above paragraph, our objective will be to find n and r such that (Ebeling, 1997):

$$\Pr[N_f \leq r | R(t) = R_0] = 1-\alpha \tag{10.3}$$

$$\Pr[N_f \leq r | R(t) = R_1(t)] = \beta \tag{10.4}$$

Equations (10.1) and (10.2) give the probability of accepting the test with high and low probability respectively. The random variable of number of failures, N_f, follows a binomial distribution and equations (10.3) and (10.4) can be restated as:

$$\sum_{i=0}^{r} \binom{n}{i} (1-R_0)^i R_0{}^{n-i} = 1-\alpha \tag{10.5}$$

$$\sum_{i=0}^{r} \binom{n}{i} (1-R_1)^i R_1^{\,n-i} = \beta \tag{10.6}$$

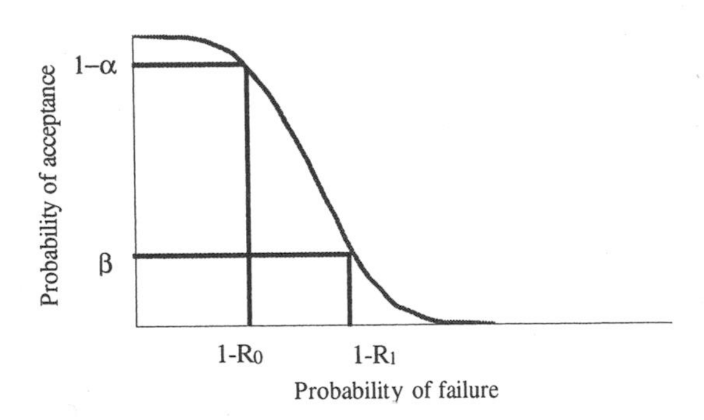

Figure 10-1. Operating characteristic curve as a function of defective probability

Once the values of R_0, R_1, α, and β, are known, the above equations can be used to find the values of n and r. One can also construct relationship between probability of failure and the probability of acceptance; the resulting curve is called operating characteristic curve (OC curve). A typical OC curve is shown in Figure 10.1. OC curve decreases monotonically as the defective probability increases.

10.2.1.2 Demonstration of constant failure rate or MTBF

If the failure rate is constant then the time to failure distribution of the system follows exponential distribution. If the time to failure is exponential, then the process of failure and the number of failures can be modeled using non-homogeneous Poisson process (NHPP). The results from the previous section can be now easily extended to demonstrate failure λ or MTBF ($1/\lambda$). The procedure for demonstrating MTBF is as follows:

Let $MTBF_0$ be the specified MTBF and $MTBF_1$ be the minimum acceptable MTBF. If the true MTBF is greater than $MTBF_0$ then the product should be accepted with a probability $1 - \alpha$ and if the true MTBF is less than $MTBF_1$ then the product should be rejected with probability $1 - \beta$. The acceptance plan using Poisson process is:

1. Determine the smallest integer c and the value of T (operating time) for which:

$$\sum_{i=0}^{c} \frac{\left(T \big/ MTBF_0 \right)^i}{i!} e^{-(T/MTBF_0)} \geq 1-\alpha \tag{10.7}$$

and

$$\sum_{i=0}^{c} \frac{\left(T \big/ MTBF_1 \right)^i}{i!} e^{-(T/MTBF_1)} \leq \beta \tag{10.8}$$

2. A test is performed for a total cumulative period of T to determine the number of failures k. The null hypothesis is rejected if k > c and accepted if k is less than or equal to c.

10.2.2 Reliability Growth Program

There are two types of product developers in this world, those who never achieve the target reliability and the rest who finally get there after several iterations of design and manufacturing process improvements. The main objective of reliability growth program is to improve a product's reliability by removing design and manufacturing weaknesses within the product.

This is a test conducted specifically to measure improvements in reliability by finding and fixing deficiencies. This test is based on what is called Test Analyze and Fix (TAAF) cycle. A growth test provides an estimate of the current product reliability. The growth test is normally conducted on new products or products which have gone through a considerable design change. A dedicated growth test can prevent the delivery of unsatisfactory product to the customer. This test is conducted on prototype samples to test and time to implement design changes based on the failures during the test. These tests should precede the RDT and qualification tests. With each test analyze and fix step, the reliability is expected to improve with the implemented design changes. Sometimes the growth may appear discontinuous and there may be a decrease in the reliability due to implemented changes. Yet in general the reliability against a cumulative test time is expected to show a definite increasing trend.

The objectives of reliability growth testing are the following:

- To predict reliability at some future time

- To estimate reliability for demonstration purpose or to estimate testing time

There are several models available for building a growth model for the data collected, of which Duane model is the one most commonly used.

10.2.2.1Duane Model

Duane developed a simple model, which is adequate for most cases of reliability growth. The total cumulative Mean time between failures (MTBF) is related to cumulative operating time T by log-log relationship:

$$\ln(M_c) = \alpha \ln(T) + \beta \qquad (10.9)$$

Where M_c is the cumulative MTBF and is given by:

$$M_c = \left(\frac{T}{n}\right)$$

T is the cumulative time of testing, n is the number of failures by time T, α and β are constants (slope and intercept of the line in equation (10.9)). The constant α is also called growth factor and is a measure of the growth programme. The value lies between 0 and 1, although 1 can never be achieved in practice. O'Connor (2004) gives comments on the efficiency of a programme as a function of α as shown in Table (10.1). This could be used as a guidance to interpret the growth test results.

Table 10-1. Efficiency of a reliability growth programme

Range of α	Efficiency of the programme
0.4 - 0.6	Programme dedicated to the removal of design weakness and to reliability
0.3 – 0.4	Well managed programme with reliability as a high priority
0.2 – 0.3	Corrective action taken for important failure mode only
< 0.2	Reliability has low priority

EXAMPLE 10.1

Table 10.2 shows the cumulative time after the start of a reliability growth programme at which the failures were observed. The test was conducted over three prototypes. The first 50 hours may have been on one prototype, or 25 hours on two prototypes, or say 20 hrs on one, 20 hours on the second and 10 hrs on third. The second failure occurred just after an hour. The third failure occurred after a cumulative hour of 120 hrs and 69 hours from the second failure and so on. Obtain the Duane plot for the following data.

Table 10-2. Reliability growth program data

Failure number (n)	Cumulative Time (T)	Cumulative MTBF Mc = (T/n)
1	50	50
2	51	25.5
3	120	40
4	150	37.5
5	250	50
6	290	48.3
7	400	57.1
8	480	60
9	630	70
10	830	83

Using least square regression, the parameters α and β in equation 10.9 are estimated by taking the natural logarithm of M_c and T in the above table. For the data given in table 10.2, we find $\alpha = 0.28$ and $\beta = 2.38$. The Duane plot of the data is given in Figure 10.2.

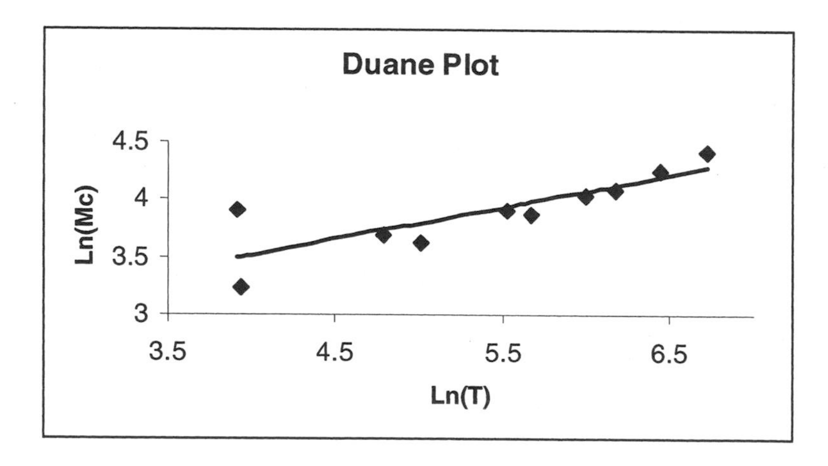

Figure 10-2. Duane plot

This value of α can be interpreted as, that the corrective actions were taken for important failure modes only. The equipments actual MTBF is known as the instantaneous MTBF and denoted by M_i and is given by:

$$M_i = \frac{M_c}{(1-\alpha)} \qquad (10.10)$$

For the given example above $M_i = \frac{83}{(1-0.28)} = 115.76$

Now using the equation 10.9 we can estimate the following:

1. Supposing the growth programme is continued up to 2000 hours, what will be the instantaneous MTBF.
2. If the required MTBF is 200 hours then how much more development is needed to reach this value.

1. Using the equation (10.9) with the estimated parameters, we have

$\text{Ln}(M_c) = 0.28\,(\text{Ln}(T)) + 2.38$

Substituting T= 2000 in this equation we get $M_c = 92.15$ hours and hence $M_i = 128.53$ hours

2. For the given MTBF 200 hours obtain the cumulative MTBF using equation (10.10).

$M_c = 200\,(1- 0.28) = 143.41$

Substituting M_c in equation in equation (10.9) and solving for T,

$\text{Ln}(M_c) = 0.28\text{Ln}(T) + 2.38$

We obtain T = 9743.9 hours. Thus (9743.9 – 830) = 8913.9 hours of testing is required.

10.3 LIFE CYCLE COST AND TOTAL COST OF OWNERSHIP

Life Cycle Cost (LCC) and the Total Cost of Ownership (TCO) are two important financial measures that are used to measure total effectiveness of a product and procurement decision making. Life cycle cost refers to all costs associated with the product or system as applied to a defined life cycle. That is, starting from requirement analysis, design, production, operation and maintenance till disposal. Total cost of ownership (TCO) is aimed at understanding the true cost of buying a particular product or service from a particular supplier. From its origins in defence equipment procurement in the US in early 1960s, the use of life cycle cost and the total cost of ownership has extended to other areas of the public and private sectors. LCC and TCO are being used to assist in decision-making, budget planning, cost control, and range of other activities that occur over the life of complex technological equipment.

LCC analysis is applied routinely to military projects. In the military sector the consumer, by funding the project and operating the related product essentially bears the total life cycle cost covering the major cost elements in all stages of a product's life cycle. It is reported by the US Department of Defence that 70% of weapon system life cycle cost is committed by the end of concept studies, 85% by the end of system definition and 95% by the end of full scale development. The term LCC analysis is rarely used in the commercial sector. Instead, the main focus is on TCO where related costs covering acquisition (purchase or lease), operation, maintenance and support are borne by the customer. In addition, the customer can also incur costs when the product is not available for use, that is, '*down time costs*'. In this book we mainly focus on the total cost of ownership.

Evaluation of a products life cycle cost involves the following three fundamental steps:

1. Development of cost breakdown structure (CBS).
2. Derivation of cost estimation relationships (CER).
3. Discounting and inflation.

Cost breakdown structure (CBS) provides foundation for any LCC analysis. The main task of CBS is to identify all relevant cost elements. CBS can easily get very complex depending on the system and the decision makers' objectives for using the LCC. If the objective is to choose the best alternative from several alternatives, then only the costs which really differentiate the alternatives should be considered. For example, one of the main tasks while procuring the weapon systems is the trial and evaluation. However, costs expended on trial and evaluation should not be included in the LCC analysis. Many cost breakdown structure exist in the literature

(Kaufmann, 1970; Fabrycky and Blanchard, 1985). The main thing to note while developing CBS is that one should not get too ambitious and try to keep the CBS simple and precise. In the next section, we discuss a framework for predicting the total cost of ownership using KISH (Keep it Simple Honey) philosophy.

Once the CBS is in place, it is necessary to calculate relevant costs. Cost estimating relationships are developed to calculate various costs. CERs are usually derived using historical data. Discounting and inflation takes care the fact that money has time value and calculates all costs to the present value for easy comparison.

10.4 TOTAL COST OF OWNERSHIP MODEL

Total cost of ownership is driven by reliability, maintainability and supportability. The objective of total cost of ownership is to minimize TCO by optimizing reliability, maintainability and supportability. Figure 10.3 illustrates the relationship between the system operational effectiveness and other design parameters (Dinesh Kumar 2000). Total cost of ownership will decrease as the reliability increases. Similarly better maintainability and supportability would decrease the maintenance and support cost and hence the total cost of ownership. However, increasing reliability, maintainability and supportability may require additional resources during the design and product development stages and hence likely to increase the initial procurement cost.

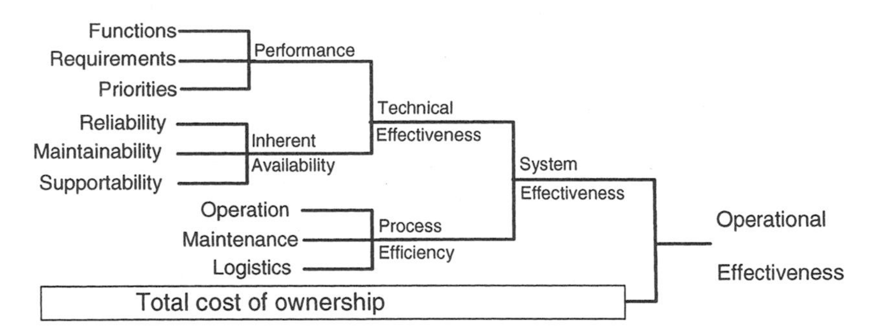

Figure 10-3. Cause and effect dependency between operational effectiveness, total cost of ownership and other design parameters

The framework for calculating total cost of ownership can be very complex depending on the procurement and asset management strategies used by the user. In this paper, we mainly focus on procurement, operation, maintenance and disposal cost, which are more relevant for assets like wagons. The framework shown in Figure 10.4 is used for evaluation of total cost of ownership.

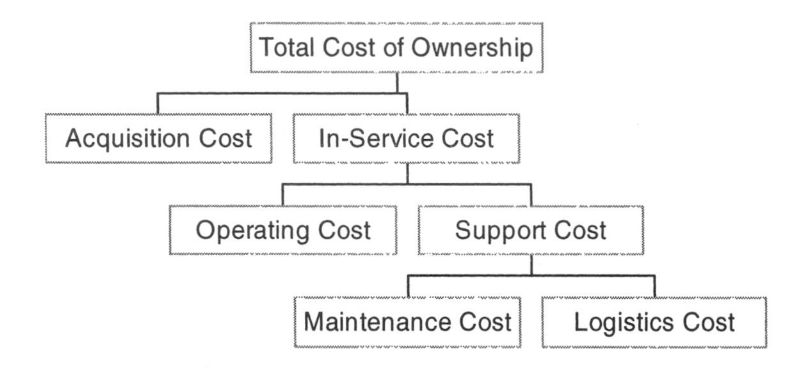

Figure 10-4. Framework for calculation of total cost of ownership

10.4.1 Mathematical models for estimation of total cost of ownership

In this section we develop mathematical models for estimation of various cost elements in the total cost of ownership. The main focus is on estimation of in-service cost. Since all the cost elements in the total cost of ownership need to be discounted to their present value, all the costs models explained in the subsequent sections are calculated on annual basis and finally discounted using appropriate discount rate.

10.4.1.1 Estimation of operating cost

The operating cost can be divided into two categories, direct operating cost and overhead cost. The direct operating cost is determined by the resources, which are required for operating the assets. The main resources for most of the system are energy consumed by the assets and the manpower required to operate the assets. The energy consumed by the assets will depend on the operational availability of the system calculated on annual basis. The operational availability, A_o, of any asset is given by:

$$A_O = \frac{MTBM}{MTBM + DT} \tag{10.11}$$

Where:

MTBM = Mean time between maintenance
DT = Down Time
Mean time between maintenance for duration, T, is given by:

$$MTBM = \frac{T}{M(T) + \frac{T}{T_{sm}}} \tag{10.12}$$

Where, M(T), is the number of failures resulting in unscheduled maintenance and T_{sm} is the time between scheduled maintenance. The down time, DT, can be estimated using the following equation:

$$DT = \frac{M(T) \times MCMT + \frac{T}{T_{sm}} \times MPMT}{M(T) + \frac{T}{T_{sm}}} \tag{10.13}$$

Where,
MCMT = Mean corrective maintenance time.
MPMT = Mean preventive maintenance time.

The number of failures resulting in unscheduled maintenance can be evaluated using renewal function and is given by:

$$M(T) = F(T) + \int_0^T M(T - x) f(x) dx \tag{10.14}$$

Where, F(T) is the cumulative distribution of the time-to-failure random variable and f(x) is the corresponding probability density function. Equation (10.14) is valid only when the failed units are replaced or when the repair is as good as new. However, in case of minimal repair and imperfect repair, one may have to use models based on non-homogeneous Poisson process or modified renewal process. For more details, the readers may refer to Ross (2000). If we assume that the energy cost and manpower cost per unit time is C_{ou} and the annual usage of an asset is T life unit, then the annual operating cost is given by:

$$C_O = A_O \times T \times C_{ou} \tag{10.15}$$

Assume that 'r' denotes the discount rate. Then the present value of the operating cost for nth period (nth year), $C_{O,n}$, is given by:

$$C_{O,n} = \frac{A_O \times T \times C_{ou}}{(1+r)^n} \tag{10.16}$$

10.4.1.2 Estimation of Maintenance Cost

The main components of maintenance cost are corrective maintenance costs, preventive maintenance costs and overhaul costs. For large systems, maintenance is carried out at several echelons (usually three echelons, namely, line maintenance, base maintenance and depot maintenance). The mathematical model developed in this section assumes there is only one maintenance echelon, which can be easily extended to multi-echelon case. Maintenance resources which were used in performing that particular maintenance task drive these costs. The maintenance cost, C_M, can be estimated using the following equation:

$$C_M = M(T) \times C_{cm} + \frac{T}{T_{sm}} \times C_{pm} + \sum_{i=1}^{k} \delta_{i,n} \times C_{OH,i} \tag{10.17}$$

Where:

C_{cm} = Average cost of corrective maintenance.
C_{pm} = Average cost of preventive maintenance

$$\delta_{i,n} = \begin{cases} 1, \text{if overhaul of type i is carried out during period n} \\ 0, \text{otherwise} \end{cases}$$

$C_{OH,i}$ represents the average cost of overhaul of type i. This cost will be added to the maintenance cost, if the type i overhaul is carried out during period n. The maintenance cost for period 'n' is given by:

$$C_{M,n} = \frac{1}{(1+r)^n}\left(M(T)\times C_{cm} + \frac{T}{T_{sm}}\times C_{pm} + \sum_{i=1}^{k} \delta_{i,n} C_{OH,i} \right) \quad (10.18)$$

10.4.1.3 Estimation of logistic support cost

Logistic support cost covers the costs associated with maintaining spare parts, maintenance facilities, test equipment and other logistics costs such as transportation costs. The spare parts contribute significant portion of the total support cost. The number of spares stocked also plays a crucial role in the operational availability of the system. Practitioners decide on the number of spare parts to be purchased based on the target fill rate, α (probability that a demand for a particular spare part can be achieved from the available stock). Usually the target fill rate is 85%. Assume that N_s represent the minimum number of spares that should be stocked to achieve a target fill rate α. Then the value of N_s can be calculated using the following equation:

$$\sum_{k=0}^{N_S} \frac{\exp(-\lambda T)\times(\lambda T)^k}{k!} \quad (10.19)$$

The above equation is valid only when the time-to-failure distribution follows exponential distribution, where λ is the failure rate. When the time-to-failure distribution is other than exponential, then we need to use renewal function to find the value of N_s to achieve the target availability. The annual logistics cost for the period n, $C_{L,n}$ is given by:

$$C_{L,n} = \frac{1}{(1+r)^n} N_s \times C_s \quad (10.20)$$

10.4.1.4 Total cost of ownership

The total cost of ownership is obtained by adding the components given by equations (10.16), (10.18) and (10.20) over the designed life of the asset. If the designed life of the asset is D, then the total cost of ownership, TCO_D, is given by:

$$TCO_D = C_P + \sum_{n=1}^{D} [C_{O,n} + C_{M,n} + C_{L,n}] + C_{MF} \quad (10.21)$$

Where C_P is the procurement price of wagon and C_{MF} is the one-time expenses of maintenance and support equipment.

10.5 TOTAL COST OF OWNERSHIP – CASE STUDY ON FREIGHT CARS ON AN ASIAN RAILWAY SYSTEM

In this case, we calculate the cost of ownership of freight cars (wagon) used in railways. These freight cars are mainly used for carrying bulk material such as iron ore, coal etc and are fitted with bogies. Bogies are the critical subsystems of the wagon (also referred to as truck) as they minimize the vibrations of the freight car while in motion. The bogie consists of two cast side frames and a floating bolster. The bolster is supported on the side frames through two groups of springs, which also incorporate the load proportional friction damping. The side frames of the bogie are connected by a fabricated mild steel spring plank suspension to maintain the bogie squareness during motion. The bogie assembly consists of the following components:

1. Wheel set with cylindrical roller bearing or wheel set with cartridge bearing.
2. Axle box/adapter, retainer bolt & side frame key assembly.
3. Side frame with friction wear plates.
4. Bolster with wear liners.
5. Spring plank, fit bolts & rivets.
6. Load bearing springs and snubber springs.
7. Friction shoe wedge.
8. Centre Pivot arrangement comprising of centre pivot, centre pivot bottom, centre pivot pin, centre pivot retainer & locking arrangement.
9. Side bearers.
10. Rubber pads on adapter.
11. Bogie brake gear.
12. Brake beam.

10.5.1 Reliability and maintainability of freight car

Railways classify the wagon failures into the following three categories:

1. Vital – causing line failure.
2. Essential – causing delay to traffic.
3. Non-essential – causing no disturbance to traffic.

The above classification enables the railway to focus c
essential components and to study their reliability and maint
service and take adequate steps to improve their performance by modification or re-design. The following three types of maintenance are practiced for wagons:

1. **Preventive maintenance Schedule (PM):** Preventive maintenance is carried out after every 6000 Km (approximately 15 days).
2. **Medium overhaul (MOH):** Medium overhaul is carried out after every 24 months. During medium overhaul, the bogie is dismantled and the wheels are de-wheeled.
3. **Major Overhaul (MaOH):** The major overhaul is carried out after every 48 months and involves complete overhaul of the wagon. However, the first major overhaul is carried out after 6 years.

For this case, we looked at the most critical components (the components that contribute towards majority of the failures). Table 10.3 shows the vital components and their time-to-failure distribution along with the estimated parameters. The time-to-failure distribution is derived using hypothetical data. The objective here is to illustrate the total cost of ownership model developed in the previous section.

Table 10-3. Vital components of freight cars and their time-to-failure distribution (λ is the failure rate, η is the scale parameter and β is the shape parameter)

S. No.	Component	Time-to-Failure Distribution	Parameters values
1.	Wheel	Weibull	η= 52, 0000 Km, $\beta = 4$
2.	Roller Bearing	Weibull	η= 250, 0000 Km, $\beta = 3$
3.	Brake Beam	Weibull	η= 160, 0000 Km, $\beta = 4$
4.	Brake Shoe	Weibull	η= 140, 0000 Km, $\beta = 3$
5.	CBC	Weibull	η= 70, 0000 Km, $\beta = 3.5$
6.	Panel Hatch	Weibull	η= 38, 0000 Km, $\beta = 4.2$
7.	Air Brake	Weibull	η= 48, 0000 Km, $\beta = 3.5$
8.	Wagon Door	Exponential	$\lambda = 6.6 \times 10^{-6}$
9.	Centre Pivot	Weibull	η= 55, 0000 Km, $\beta = 3.8$

All critical components except freight car door follow Weibull distribution. The time-to-failure distribution of the wagon door is exponential, since most of the wagon door failures are caused due to mishandling. The time-to-failure of the freight car follows an exponential distribution with mean time between failures of 16000 Km.

10.5.2 Operational availability of freight car

All the life units are measured in terms of kilometers and thus the PM, medium overhaul and major overhaul are converted to Km. The preventive maintenance interval is approximately 15 days, that is after every 6000 Km, and during PM, the wagon is out of service for 2 days (that is 800 Km). Whenever, the wagon requires corrective maintenance, it is likely to be out of service for 4 days (that is 1600 Km). The usage of the wagon for every month is 12,000 Km. Using these data, the mean time between maintenance in one year, (144,000 Km) is given by:

$$MTBM_{wagon} = \frac{144000}{M(144000) + \dfrac{144000}{6000}} = 4363Km \quad (10.22)$$

The Down time is given by:

$$DT_{wagon} = \frac{M(T) \times MCMT + (\dfrac{T}{T_{sm}}) \times MPMT}{M(T) + \dfrac{T}{T_{sm}}} \quad (10.23)$$

$$= \frac{9 \times 1600 + 24 \times 800}{9 + 24} = 1018km$$

Using equations (10.22) and (10.23), we get the operational availability of the wagon as:

$$A_{wagon} = \frac{MTBM_{wagon}}{MTBM_{wagon} + DT_{wagon}} = \frac{4363}{4363 + 1018} = 0.8108 \quad (10.24)$$

Thus, the operational availability of the wagon is 81.08%.

10.5.3 Operating cost of freight car

The operational availability value can now be used to calculate the operating cost of the freight car. For the sake of mathematical simplicity, we calculate the cost of ownership for 6 years from commissioning of the wagon. The life of a wagon can be as high as 35 years. However, to cut lot of mathematics and huge tables, we calculate total cost of ownership for the first 6 years. The reason for choosing 6 year period is that the major overhaul of the wagon happens after 6 years. Assume:

C_{ou} = \$1 per Km

Then, the operating cost for first six years, at an interest rate of 6% is given in the following table (table 10.4):

Table 10-4. Present value of the operating cost

Year	PV of the operating cost (in \$)
1	9178.868
2	8659.309
3	8169.16
4	7706.755
5	7270.523
6	6858.984
Total	54314.34

10.5.4 Maintenance cost for freight car

The cost of maintenance for six years can be calculated using the equation (10.18). We make the following assumptions:

C_{cm} = Cost of corrective maintenance = \$800
C_{pm} = Cost of preventive maintenance = \$1500
$C_{OH,1}$ = Cost of medium overhaul = \$8000
$C_{OH,2}$ = Cost of major overhaul = \$15000

Table 10.5, shows the present value of the maintenance cost for the first six years.

Table 10-5. Present value of the maintenance cost

Year	PV of the logistics cost (in \$)
1	30849.06
2	36222.86
3	27455.55
4	37782.87
5	24435.34
6	28691.89
Total	30849.06

In table 10.5, one can notice, cost fluctuation during year 2, 4 and 6. This is due to minor and major overhaul carried out during that period.

10.5.5 Logistic support cost

The logistics support cost can be estimated using equation (10.20). This involves the use of renewal process to estimate the spares requirement for each of the components shown in table 10.3. Assuming, N_s x C_s = \$ 3000, the present value of the logistics cost for six years is shown in Table 10.6.

Table 10-6. Present value of the logistic support cost

Year	PV of the logistics cost (in \$)
1	2830.189
2	2669.989
3	2518.858
4	2376.281
5	2241.775
6	2114.882
Total	14751.97

10.5.6 Total cost of ownership of freight car

The total cost of ownership of freight car for the first 6 years is obtained by adding the components given by equations (10.16), (10.18) and (10.20). Assume that C_{MF} = \$20000. The total cost of ownership, TCO_D, for six years, is given by:

$$TCO_D = C_P + \sum_{n=1}^{D} [C_{O,n} + C_{M,n} + C_{L,n}] + C_{MF}$$

$$= 40000 + 54314.34 + 185437.6 + 14751.97 + 20000$$

$$= \$314502.6$$

The cost of ownership is calculated for six years, which is the first major overhauling period. The above cost can be divided by the duration, to calculate TCO per year, which then can be used for comparing different designs alternatives.

10.6 IMPACT OF SIGMA LEVEL ON TOTAL COST OF OWNERSHIP

One of the main reasons and incentives for organizations trying to achieve higher sigma levels is the reduction in the life cycle cost due to poor quality. As the sigma level increases, the cost of poor quality decreases. As far as users are concerned, higher sigma levels should mean lower cost of ownership. This is due to the fact that a higher sigma level implies lesser deviation from the target reliability. Assume that σ_A and σ_B denote the sigma level of two processes A and B respectively. The corresponding DPMO for processes A and B are given by:

$$DPMO(\sigma_A) = \exp\left(\frac{29.37 - (\sigma_A - 0.8406)^2}{2.221}\right) \tag{10.25}$$

$$DPMO(\sigma_B) = \exp\left(\frac{29.37 - (\sigma_B - 0.8406)^2}{2.221}\right) \tag{10.26}$$

Equations (10.25) and (10.26) are derived from equation (3.12) in chapter 3 (we will ignore the error part of this equation). The difference between DPMO(σ_A) and DPMO(σ_B) will have significant impact on the maintenance cost of the system and thus on the cost of ownership. In addition to the maintenance cost, the user may be forced to maintain a larger fleet to maintain certain level of operational availability in case of lower sigma level resulting in higher total cost of ownership.

From equations (10.25) and (10.26) it is evident that the decrease in cost of poor quality decreases as the sigma value increases. That is a company which increases its sigma level from 3 sigma to 4 sigma achieves higher savings compared to a company which increases its sigma level from 5 sigma to 6 sigma. So it is natural for the management to question whether it is worth trying to achieve a higher sigma level, if the company is already on the higher side, say above 5 sigma level. The answer to this very valid question lies in the fact of how the company uses Six Sigma. If Six Sigma is used to resolve operational issues or as an operational strategy then the benefit of increasing the sigma level decreases as the sigma level increases. However, if Six Sigma is used as a corporate strategy, the cost benefits are much higher. The Design for Six Sigma (DFSS) concept is developed as a corporate strategy for new product development (see section 10 of this chapter).

EXAMPLE 10.2

The **G** Company and the **P** Company are the major manufacturers of jet engines. The reliability of jet engines is usually measured using number of removal per 1000 flying hours. Both these companies have quoted same removal rate for their engines to be fitted in a new aircraft developed by **B** Company. Each additional removal of the engine from the aircraft other than the planned removal will result in $50000 to the airlines. If the Sigma level of company **G** and **P** are 4.85 and 4.8 respectively, calculate the additional amount of money which may be incurred by the airlines that chooses engines manufactured by P Company.

Using equations (10.25) and (10.26) the corresponding DPMOs are given by:

DPMO (4.85) $\approx$ 398 and DPMO (4.8) $\approx$ 476

The airlines which buy engines from P Company would face 78 additional removals compared to the airlines that buy engines from G Company. The corresponding cost is $ 3.9 million.

10.7 SIX SIGMA PROJECT MANAGEMENT

Six Sigma is a project based process improvement strategy. Selection of projects plays a critical role in successful implementation of Six Sigma. Like any process improvement initiative, Six Sigma does not guarantee success for all projects. The selected project should allow the Six Sigma team to implement DMAIC methodology within the schedule and cost. Any deviation from the schedule and cost is likely to reduce the benefit.

Traditional operations research techniques can be used for selection of Six Sigma projects. Selection of projects from several alternatives is basically a multi-criteria multi-person (MCMP) optimization problem. Operations research techniques such as analytic hierarchy process (AHP), Goal Programming (GP) and Theory of Constraints (TOC) can be used to select Six Sigma project. In this section we illustrate how AHP can be used to select Six Sigma projects.

10.7.1 Analytic hierarchy process (AHP) for Six Sigma project selection

AHP is a multi-criteria decision making technique which can be used to choose best alternative among a number of alternatives (Saaty, 1980). AHP methodology uses the following steps to identify the best alternative from a set of alternatives:

1. Identify the overall goal of the problem.
2. Identify criteria that must be satisfied by the overall goal.
3. Identify sub-criteria under each criterion if applicable.
4. Identify stakeholders.
5. Construct pair-wise comparison matrix based on the feedback from all the stakeholders.
6. Rank the alternatives from the most preferred to the least preferred by analyzing the pair-wise comparison matrices.
7. Choose the most preferred one from step 6 for implementation of Six Sigma. In case the management decides to implement Six Sigma in more than one project, then the rank list developed in step 6 can be used accordingly.

The following are some of the criteria for Six Sigma project selection:

1. Scope of the project.
2. Clearly defined deliverables
3. Project duration and cost
4. Availability of data required for DMAIC cycle
5. Availability of resources
6. Project risk

Once the criteria are identified and the problem is arranged in a hierarchical manner, AHP develops pair-wise comparison matrix which is the building block of AHP methodology. AHP methodology is based on the following two axioms:

- If P_c (A, B) is the paired comparison of alternatives under criteria C, the P_c (B, A) = 1/ P_c (A, B). This is called reciprocal axiom.
- The alternatives being compared should not differ by too much in the criteria being compared. This is called homogeneity axiom.

A pair-wise comparison matrix is a $n \times n$ matrix with values a_{ij} indicating the value of alternative i relative to alternative j. A nine point scale is used by AHP for making pair wise comparisons (Table 10.7).

Calculation of Priorities

The priority list (weights) of alternatives under various criteria is calculated by normalizing the pair-wise comparison matrix. The steps involved in calculating the priority list are:

1. Calculate the column sum of each column in the pair-wise comparison matrix.
2. Divide the elements of pair-wise comparison matrix by its column sum, the resulting matrix is called normalized pair-wise comparison matrix.
3. Calculate the average of row elements of each row in the pair-wise comparison matrix. The resulting vector gives the priority list of various alternatives under a particular criterion.
4. Repeat steps 1-3 for all the criteria.

Table 10-7. Nine point scale used by AHP to construct pairwise comparison matrix

Intensity of Importance	Definition
1	Equal importance
3	Weak importance of one over other
5	Strong Importance
7	Demonstrated Importance
9	Absolute Importance
2,4,6,8	Intermediate Values
Reciprocals of the above	If activity *i* has one of the above numbers assigned to it when compared with activity *j*, then *j* has the reciprocal value when compared with *i*.
1.1 – 1.9	When elements are close and nearly indistinguishable

Not all criteria will be treated equally important. The procedure for finding priority list for alternatives under various criteria is repeated to find the weighted priority list of the criteria. Assume that there are m criteria, and n alternatives. Steps 1-3 are used to calculate the weights w_i of the criteria i. For each criteria i, the alternatives (n alternatives) are compared to determine their w_{ij} (j = 1, 2,..., n) with respect to criteria i. The overall priority list by considering all criteria for each alternative j is given by:

$$W_j = w_{1j}w_1 + w_{2j}w_2 + \ldots + w_{mj}w_m \tag{10.27}$$

The one with the higher weight should be chosen for implementing Six Sigma.

EXAMPLE 10.3

The management of Jindal limited is trying to choose a project to implement Six Sigma from three planned projects (projects A, B and C). The decision makers at Jindal limited have identified four criteria that should be considered while choosing the project. The four criteria for project selection are listed below:

1. Project cost and duration
2. Data availability
3. Increase in the Sigma level.
4. Benefit after implementation of Six Sigma

The decision makers have derived the pair-wise comparison matrix for all the alternatives (projects A, B and C) under each of the four criteria, which are shown along with the pair-wise comparison matrix for all the four criteria.

Pair-wise comparison matrix for criteria 1

	A	B	C
A	1	2	8
B	1/2	1	6
C	1/8	1/6	1

Pair-wise comparison matrix for criteria 2

	A	B	C
A	1	1/3	¼
B	3	1	½
C	4	2	1

Pair-wise comparison matrix for criteria 3

	A	B	C
A	1	¼	1/6
B	4	1	1/3
C	6	3	1

Pair-wise comparison matrix for criteria 4

	A	B	C
A	1	1/3	4
B	3	1	7
C	1/4	1/7	1

The pair-wise comparison matrix for four criteria:

	Criteria 1	Criteria 2	Criteria 3	Criteria 4
Criteria 1	1	3	2	2
Criteria 2	1/3	1	¼	¼
Criteria 3	½	4	1	½
Criteria 4	1/2	4	2	1

The priority list for alternatives A, B and C under criteria 1 can be calculated using steps 1-3 of AHP procedure.

	A	B	C
A	1	2	8
B	1/2	1	6
C	1/8	1/6	1
Column sum	13/8	19/6	15

The normalized pair-wise comparison matrix (dividing elements by its column sum) is given by:

	A	B	C
A	0.615	0.632	0.533
B	0.308	0.316	0.400
C	0.077	0.053	0.067

The corresponding priority vector (average of row elements) is given by:

A	0.593
B	0.341
C	0.066

By repeating the procedure for alternatives under other criteria, we get:

Criteria 2	
A	0.123
B	0.320
C	0.557

Criteria 3	
A	0.087
B	0.274
C	0.639

Criteria 4	
A	0.265
B	0.655
C	0.080

Priority weights for four criteria are given by:

Criteria 1	0.398
Criteria 2	0.085
Criteria 3	0.218
Criteria 4	0.299

Now using equation (10.27), the overall weight for projects A, B and C are given by:

Overall weight	
A	0.265
B	0.421
C	0.314

Thus project B should be chosen for implementation of Six Sigma. If the management decides to implement Six Sigma in two projects then projects B and C should be chosen.

10.8 DMAIC CYCLE

Define-Measure-Analyze-Improve-Control (DMAIC) methodology provides a roadmap to conduct Six-Sigma projects and often used as a "gated process" for controlling various phases in a project. DMAIC has become such an integral part of Six-Sigma that the implementation of Six-Sigma almost always follows a DMAIC framework in any company. DMAIC uses several mathematical, statistical, engineering and management tools to improve the processes in a sequential way. Although it is a sequential process, the project team may go through several iterations until all project objectives are met (Figure 10.5). Below, we give a theoretical definition of each phase of DMAIC at first and then describe what it transforms to while implementing a specific project, followed by specific Six-Sigma tools that are used in each phase. A case study on locomotive starter batteries is used to illustrate how the DMAIC methodology may be used in real life projects.

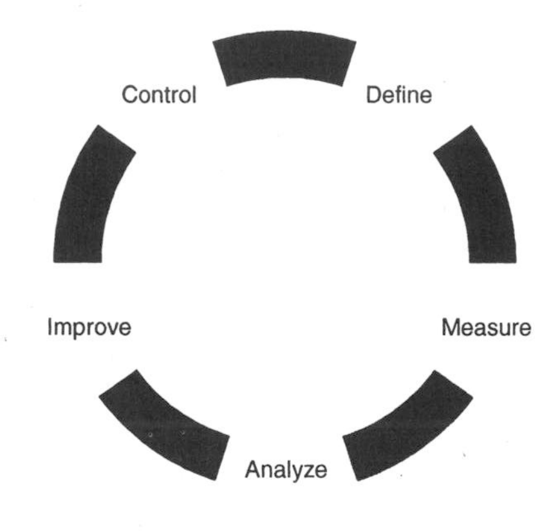

Figure 10-5. DMAIC Cycle

10.8.1 Define phase

Define is the starting phase of the DMAIC methodology. During the Define phase the project team defines the project scope, project objectives or goals. The project objectives should be in alignment with customer requirements. The goals could be the organizational goal at a strategic level, or the goal to increase the output at the operational level or the goal to decrease number of defects at a project level or the goal to meet a specific customer requirement. It is important to define the goals by directly communicating to all the stake holders and keeping their benefits in view.

In the *define* phase, typically a problem is identified, a project team is formed and given the responsibility to solve the problem, along with the required resources. The activities during the Define phase are:

1. Identify the customers
2. Identify what is important to the customer (Voice of Customers).
3. Identify Critical to Quality (CTQ) parameters.
4. Identify problems.
5. Identify improvement opportunities.

6. Set targets for the project.

Table 10.8 gives the some of the tools that can be used during the Define phase and their purpose.

Table 10-8. Commonly used tools during the Define stage

Tool	Purpose
Quality function Deployment	To identify what is important to customer
CTQ tree	To identify Critical to Quality Parameters
Analytic Hierarchy Process	To prioritize customer requirements
SIPOC map (Suppliers, inputs, processes, outputs and customers map)	To identify the scope of the project
Failure Modes Effects and Criticality Analysis	To identify potential problems and their impact on the customers

10.8.2 Measure Phase

During the Measure phase of the DMAIC methodology various metrics to measure the performance of the process need to be established. The metrics should reflect the customer requirements and critical to quality parameters. Once the metrics are identified the next step is to collect the relevant data. The type of data that is needed and the corresponding metrics are identified in order to progress towards the goal. The main objective of the Measure phase is to collect the data pertaining to the problem, and measuring the performance of the process through appropriate measures. Frequently used measures of performance are variability in the process, DPMO, Yield, probability of failure free operation, mean (operating) time between failures, operational availability, cycle time, cost of poor quality, total cost of ownership etc. The main activities during the define phase are:

1. Identify appropriate metrics to measure the current processes.
2. Collect the relevant data required to quantify the process measures.
3. Analyze the data and evaluate the performance of the processes and other critical to quality parameters (CTQs).

10.8.3 Analyze Phase

Analyze why the current process is not able to meet the customer requirement and generate solutions to fix the problem. In other words, during the analyze phase problems are identified and solutions are generated to achieve the goals set during the Define stage of the DMAIC methodology. If a product is not meeting the customer requirements, it is necessary to identify the processes that are responsible for the failure to meet customer requirements. The data collected from the Measure phase is used to identify the process weaknesses. Once the process weaknesses are known the next step is to generate solutions to remove the weaknesses. In the analyze phase, one gets to the bottom of the root causes by testing out theories as to what might have caused the problem. The main focus during this stage is the input factors and processes that drive the output. The following activities are routinely carried out during the Analyze stage.

1. Identify the problems in the input and the processes.
2. Assess the impact of input and process related problems on the customer requirement and critical to quality parameters.
3. Prioritize the problems.
4. Identify root causes.
5. Generate solutions to eliminate input and process related problems.

Table 10-9. Tools for Analyze stage

Tool	Purpose
Root cause analysis	To identify the main cause of the problem
Data mining	Set of statistical tools used to extract useful information from a data
Design of Experiments (DoE)	To study influence of several factors
Pareto chart	Graphical representation of problems so that effort can be expended on the critical problems
Failure Modes Effects and Criticality Analysis	To identify potential problems and their impact on the customers
Cause and Effect Diagram	To identify possible causes from various sources such as material, manpower, equipment and methods.
Simulation	To duplicate the system using computer coding to learn the behaviour of the system.

Table 10.9 gives some of the tools that can be used during the Analyze stage and their purpose.

10.8.4 Improve phase

The knowledge generated so far in the Define, Measure, and Analyze stage is used to fix the problem by generating ideas. Innovative solutions from previous phase are implemented to remove the gap between the current state and desired state of the system in order to achieve the goal. The product is made competitive by increasing the performance and reducing the defects. The improvement is validated with the help of statistical tools. In general, Improve phase deals with removal of root causes through design changes in the process that was producing defects. The following activities are carried out during the Improve stage.

1. Generate ideas to solve the problem.
2. Screen the ideas (say using Pugh matrix) and choose the best course of action.
3. Test solutions using prototype or on a pilot study.
4. Implement solution.

Tools used during the Improve stage are: project planning and management tools, prototype and pilot studies, Force field diagrams

10.8.5 Control phase

The improvements are internalized and institutionalized through various controls. The improvements achieved through previous stages are consolidated and effort is made to sustain the improvement achieved. The process changes are documented and made available to the entire community within the organization. These controls used may be organizational, like incentive systems, policies etc. or operational, like procedures, operating instructions etc. The objective of the control phase is to ensure that the improvement does not fade away over time. In the control phase, improvements in the process are retained by designing controls that ensure the defects from reoccurring.

Tools such as statistical process control, process capability analysis, cost estimation models, are used during the Control stage.

10.9 CASE STUDY: DMAIC METHODOLOGY

In this section we discuss a case to illustrate how DMAIC methodology can be used in real life. The names in the case are changed to maintain the confidentiality.

10.9.1 PowerPlus battery manufacturing company

PowePlus is one of the largest manufacturers of batteries. PowerPlus batteries are classified into two major types: Automobile batteries and Industrial batteries. PowerPlus is a major supplier of Locomotive Starter batteries to railways in Asian countries. Diesel locomotive starter battery is basically a flooded lead acid battery. PowerPlus provides 18 month warranty for these batteries. In the recent years, the customer care department of PowerPlus received complaints from the customers that the battery life of PowerPlus batteries is much lesser compared to batteries supplied by other manufacturers and the difference was more than 12 months.

The management of PowerPlus decided to fix the problems associated with locomotive starter batteries using Six Sigma and formed a team of 7 members with 2 Black Belts and 5 Green Belts with sufficient skills in statistical techniques as well as design and manufacturing of lead acid batteries. The details about lead-acid batteries used by railways are given below:

1. The diesel locomotive starter battery used by railways is a flooded lead acid battery of 450 Ampere Hour capacity.
2. The positive plates are manufactured from lead oxide where as lead is used for negative plates.
3. Positive and negative plates are insulated by separators. The separators are made of highly porous material which allows free movement of ions.
4. Tubular casting in gauntlet is used for positive plates (Anode) whereas a grid structure is used for construction of negative plates (cathode).
5. The electrolyte is made of sulphuric acid and water.
6. The container is made of hard rubber as well as polypropylene assembly for some models
7. The battery weighs about 50 Kgs.

In the next few sections, we describe how the team went about solving the problem using DMAIC methodology.

10.9.1.1 Define stage

It is evident from the customer complaints that the main problem associated with the battery is its life. The longer the life of the battery, the total cost of ownership per hour will be lesser. So, the main objective of this Six Sigma project is to extend the life of the battery and thus reduce the total cost of ownership.

10.9.1.2 Measure

The life of the locomotive starter batteries is measured using MTTF measured in number of months. These batteries are not repairable and are condemned and recycled to extract the lead (mandated by pollution control board) which is the costliest material used in battery manufacturing. The failure data of the 33 batteries are collected and is shown in Table 10.10.

Table 10-10. Failure data of locomotive starter batteries

41	41	41	43	47	48	51
51	52	52	52	52	52	52
52	53	57	57	59	60	61
61	61	61	61	63	67	69
71	71	71	71	78		

The average life of the batteries in months based on the data in table 10.10 is close to 57 months with a standard deviation of 9.69 months. Since the customer care department has already set a target that the battery life should be improved by at least 18 months (to catch up with the 12 months difference plus additional 6 months to become more competitive in the market), the Six Sigma project team set a goal of 75 months MTTF for the batteries that is an improvement of 18 months in MTTF. It was also decided that the standard deviation of 9.69 should be reduced to less than 4 months.

10.9.1.3 Analyze

The data is further analyzed to identify the failure modes and failure causes. The failure modes as observed by the Customers are shown using the Pareto chart in Figure 10.6. From Figure 10.6, it is evident that the majority failure modes are capacity related followed by broken container. Further analysis was carried out to learn the causes of:

1. Capacity related problems.
2. Container broken

The above to failures modes account for more than 72% of all failures and finding solutions to remove these failure modes will result in higher MTTF and thus the team will be able to achieve its target. Eliminating these failure modes would require understanding the causes of these failure modes. A further analysis has shown that the main cause of capacity related problems is the corrosion of plates. Note that these plates are immersed in sulphuric acid and any manufacturing deficiency is likely to result in corrosion. One possible reason for corrosion is the existence of impurities in the plates. However, PowerPlus used very high quality lead and the team concluded that the corrosion in this case may not be due to the impurities in the lead alloy used for manufacturing positive and negative plates. However, if the geometry of the plates is not uniform (that is voids at micro-structural level in the plates), then the acid may occupy the voids and result in corrosion.

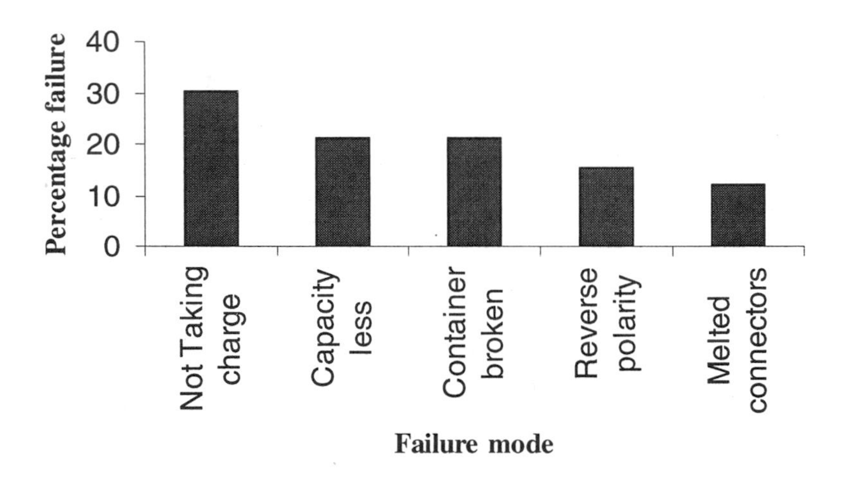

Figure 10-6. Pareto chart for failure mode of batteries

Currently the manufacturing of the plates at PowerPlus is carried out using gravity casting and thus there is a possibility of voids in the plates since the process of gravity casting is likely to result in non-uniformity in the plates. A further study using electron microscope has revealed the fact that the casting is not uniform in the plates used for construction of these batteries. Alternative process may be required to eliminate this problem.

Coming to the second failure mode, the broken container, these containers are made of hard rubber and do not break under normal operating conditions. Most of these failures may be due to mishandling of the batteries by the users. Further investigation also revealed that the batteries are just dropped from the truck to the ground at the unloading locations without proper shock absorbers. However, customer is always right, so solutions should be generated to solve this problem as well.

10.9.1.4 Improve Stage

Pressure casting is known to give better geometric structure for the plates and the team decided to change the manufacturing process of plates from gravity casting to pressure casting. A pressure casting machine was procured at the cost of $750000. This cost was justified through the increase in productivity and quality loss; further this process will reduce the voids in casting and would increase the life of the batteries.

Another improvement was made to the container alloy by changing the formulation of the container polymer. The company also took charge of the delivery of the batteries to overcome the mishandling of batteries during movement of batteries. The failure data was collected to verify the increase in the battery life.

10.9.1.5 Control

The failure data of batteries are collected and continuously checked for the average life of the batteries. Reliability life such as B_{10} life (time by which 10% of the batteries fail) are used to compare the new process with the old processes.

10.10 DESIGN FOR SIX SIGMA (DFSS)

The DMAIC methodology is used at the operational level, that is to improve the manufacturing processes and is appropriate to use from the production stage of the products life cycle. Design for Six Sigma is more focused on prevention of defects at the design stage of the product development. In DFSS, design engineers interpret and design the functionality of customer requirements by optimizing both customer needs and organizational objectives. DFSS consists of five stages, namely (i) Define (ii) Measure (iii) Analyze (iv) Design and (v) Verify, is popularly known as DMADV methodology. DMADV is appropriate for new product development where the product and processes do not exist. The activities during various stages in DFSS are given below:

- **Define**: Develop new product development (NPD) strategy, scope of the NPD project.
- **Measure**: Understand the customer requirements and Critical to Quality parameters.
- **Analyze**: Develop conceptual design after analyzing various design options.
- **Design**: Develop detailed design of the product and process
- **Verify**: Develop prototype to verify the effectiveness of the design and check whether the product has met all the goals set during the previous stages of product development.

10.11 SIX SIGMA BLACK BELT AND GREEN BELT

Implementation of Six Sigma require dedicated and skilled work force. Without sufficient knowledge about various tools and techniques it is not possible to implement. This aspect of manpower requirement is taken care by crating Black Belts and Green Belts.

Black belts are responsible for leading any Six Sigma projects. Apart from Six Sigma tools Black Belts should also possess leadership qualities. Selection of Black Belts is crucial for successful implementation of Six Sigma. Black Belts dedicate their entire time on implementing Six Sigma projects. While on Black Belt duty, they are released from their normal position and are allowed to concentrate on Six Sigma implementation.

Black belts are supported by a team of skilled workforce called Green Belts. The main difference between the black belts and green belts is that the training required for a green belt is much less that that of black belts and in addition the green belts spend only part of their working time on Six Sigma projects. There are many organizations provide that Black Belt and Green Belt training and certification.

Appendix

1. Table 1: Standard normal distribution table.
2. Table 2: t distribution table
3. Table 3: Gamma function table
4. Table 4: Sigma level, Yield and DPMO table

Table 1: Area under standard normal distribution N(0, 1)

z	Φ(z)	z	Φ(z)	z	Φ(z)	z	Φ(z)
-4	3.1E-05	-2.0	0.0227500	0	0.5	2.0	0.97724994
-3.9	4.8E-05	-1.9	0.0287164	0.1	0.5398279	2.1	0.98213564
-3.8	7.2E-05	-1.8	0.0359302	0.2	0.57925969	2.2	0.9860966
-3.7	0.00010	-1.7	0.0445654	0.3	0.61791136	2.3	0.98927592
-3.6	0.000159	-1.6	0.0547992	0.4	0.6554217	2.4	0.99180247
-3.5	0.000232	-1.5	0.0668072	0.5	0.69146247	2.5	0.99379032
-3.4	0.000336	-1.4	0.0807567	0.6	0.72574694	2.6	0.99533878
-3.3	0.000483	-1.3	0.0968005	0.7	0.75803642	2.7	0.99653298
-3.2	0.000687	-1.2	0.1150697	0.8	0.78814467	2.8	0.99744481
-3.1	0.000967	-1.1	0.1356661	0.9	0.81593991	2.9	0.99813412
-3	0.001349	-1	0.1586552	1	0.84134474	3	0.99865003
-2.9	0.001865	-0.9	0.1840600	1.1	0.8643339	3.1	0.99903233
-2.8	0.002555	-0.8	0.2118553	1.2	0.88493027	3.2	0.9993128
-2.7	0.003467	-0.7	0.2419635	1.3	0.90319945	3.3	0.99951652
-2.6	0.004661	-0.6	0.2742530	1.4	0.91924329	3.4	0.99966302
-2.5	0.006209	-0.5	0.3085375	1.5	0.93319277	3.5	0.99976733
-2.4	0.008197	-0.4	0.3445783	1.6	0.94520071	3.6	0.99984085
-2.3	0.010724	-0.3	0.3820886	1.7	0.95543457	3.7	0.99989217
-2.2	0.013903	-0.2	0.420740	1.8	0.96406973	3.8	0.99992763
-2.1	0.017864	-0.1	0.460172	1.9	0.97128351	3.9	0.99995188

Table 2: Critical t values with n degrees of freedom

	Value of α				
N	0.100	0.050	0.025	0.010	0.005
1	3.078	6.314	12.706	31.821	63.657
2	1.886	2.920	4.303	6.695	9.925
3	1.639	2.353	3.182	4.541	5.841
4	1.533	2.132	2.776	3.747	4.604
5	1.476	2.015	2.571	3.365	4.032
6	1.440	1.943	2.447	3.143	3.707
7	1.415	1.895	2.365	2.998	3.499
8	1.397	1.860	2.306	2.896	3.355
9	1.383	1.833	2.262	2.821	3.250
10	1.372	1.812	2.228	2.764	3.169
11	1.366	1.796	2.201	2.718	3.106
12	1.356	1.782	2.179	2.681	3.055
13	1.350	1.771	2.160	2.650	3.012
14	1.345	1.761	2.145	2.624	2.997
15	1.341	1.753	2.131	2.602	2.947
16	1.337	1.746	2.120	2.583	2.921
17	1.333	1.740	2.110	2.567	2.898
18	1.330	1.734	2.101	2.552	2.878
19	1.328	1.729	2.093	2.539	2.861
20	1.325	1.725	2.086	2.528	2.845
21	1.323	1.721	2.080	2.518	2.831
22	1.321	1.717	2.074	2.508	2.819
23	1.319	1.714	2.069	2.500	2.807
24	1.318	1.711	2.064	2.492	2.797
25	1.316	1.708	2.060	2.485	2.787
26	1.315	1.706	2.056	2.479	2.799
27	1.314	1.703	2.052	2.473	2.771
28	1.313	1.701	2.048	2.467	2.763
29	1.311	1.699	2.045	2.462	2.756
∞	1.282	1.645	1.960	2.326	2.576

Table 3: Gamma Function Table

x	Γ(x)	x	Γ(x)	X	Γ(x)	x	Γ(x)
0.5	1.77245	2.1	1.046486	3.7	4.170652	5.3	38.07798
0.6	1.48919	2.2	1.101802	3.8	4.694174	5.4	44.59885
0.7	1.29805	2.3	1.166712	3.9	5.29933	5.5	52.34278
0.8	1.16423	2.4	1.242169	4	6	5.6	61.55391
0.9	1.06862	2.5	1.32934	4.1	6.812623	5.7	72.52763
1	1	2.6	1.429625	4.2	7.75669	5.8	85.62174
1.1	0.95135	2.7	1.544686	4.3	8.855343	5.9	101.2702
1.2	0.91816	2.8	1.676491	4.4	10.1361	6	120
1.3	0.89747	2.9	1.827355	4.5	11.63173	6.1	142.4519
1.4	0.88726	3	2	4.6	13.38129	6.2	169.4061
1.5	0.88622	3.1	2.19762	4.7	15.43141	6.3	201.8133
1.6	0.89351	3.2	2.423965	4.8	17.83786	6.4	240.8338
1.7	0.90863	3.3	2.683437	4.9	20.66739	6.5	287.8853
1.8	0.93138	3.4	2.981206	5	24	6.6	344.7019
1.9	0.96176	3.5	3.323351	5.1	27.93175	6.7	413.4075
2	1	3.6	3.717024	5.2	32.5781	6.8	496.6061

Table 4: Sigma Level, DPMO and Yield table

PROCESS SIGMA(Z) (including the 1.5 shift)	DPMO	YIELD (Y in %)
0.0	933,000	7
0.1	919,000	8
0.2	903,000	10
0.3	885,000	12
0.4	864,000	14
0.5	841,000	16
0.6	816,000	18
0.7	788,000	21
0.8	758,000	24
0.9	726,000	27
1.0	691,000	31
1.1	655,000	34
1.2	618,000	38
1.3	579,000	42
1.4	540,000	46
1.5	500,000	50
1.6	460,000	54
1.7	421,000	57.9
1.8	382,000	61.8
1.9	345,000	65.5
2.0	309,000	69.1
2.1	274,000	72.6
2.2	242,000	75.8
2.3	212,000	78.8
2.4	184,000	81.6
2.5	159,000	84.1
2.6	136,000	86.4
2.7	115,000	88.5
2.8	96,800	90.32
2.9	80,800	91.92
3.0	66,800	93.32

Table 4 continued

PROCESS SIGMA(Z) (including the 1.5 shift)	DPMO	YIELD (Y in %)
3.1	54,800	94.52
3.2	44,600	95.54
3.3	35,900	96.41
3.4	28,700	97.13
3.5	22,800	97.72
3.6	17,900	98.21
3.7	13,900	98.61
3.8	10,700	98.93
3.9	8,200	99.18
4.0	6,210	99.379
4.1	4,660	99.534
4.2	3,470	99.653
4.3	2,560	99.744
4.4	1,870	99.813
4.5	1,350	99.865
4.6	968	99.903
4.7	687	99.931
4.8	483	99.952
4.9	337	99.966
5.0	233	99.9767
5.1	159	99.9841
5.2	108	99.9892
5.3	72	99.9928
5.4	48	99.9952
5.5	32	99.9968
5.6	21	99.9979
5.7	13	99.9987
5.8	9	99.9991
5.9	5	99.9995
6.0	3.4	99.99966

References

Abernethy, R. B. (1983), *Weibull Analysis Handbook,* AFWAL-TR-83-2079, Wright-Paterson AFB, Ohio

Abromovitz M, and Stegun, I. A. (1972), *Handbook of Mathematical Functions,* Dover, New York.

Aggarwal, K K. (1993), *Reliability Engineering,* Kluwer Academic Publishers, New York, USA

Anon (1984), Electronic Reliability Design Handbook (MIL-HDBK-388),' US Department of Defense, Washington D.C.

Ascher, H. and Feingold, H. (1984), *Repairable Systems Reliability – Modelling, Inference, Misconceptions and their causes, –* Marcel Dekker, New York.

Barlow, R.E., Proschan, F., and Hunter, L C. (1965), *Mathematical Theory of Reliability,* John Wiley and Sons, New York.

Barlow, R.E., and Proschan, F. (1975), *Statistical Theory of Reliability and Life Testing*, Holt, Rhinehart & Winston, Inc., New York.

Bazovsky, I. (1961), *Reliability Theory and Practice*, Prentice Hall, New York.

Berman, O., and Dinesh Kumar, U. (1999), Optimisation Models for Recovery Block Schemes, *European Journal of Operational Research,* 115, 368-379.

Berman, O., and Dinesh Kumar, U. (1999), Optimization Models for Complex Recovery Block Schemes, *Computers and Operations Research,* 26, 525-544.

Bertels, T (Ed), (2003), *Rath and Strong's Six Sigma Leadership Handbook*, John Wiley & Sons, New Jersey.

Bhaskar T and Dinesh Kumar, U. (2005), A Cost Model for N-Version Programming with Imperfect Debugging', Forthcoming *Journal of Operational Research Society.*

Billinton, R., and Allan, R. N. (1983), *Reliability Evaluation of Engineering Systems*, Plenum Press, NY

Birolini, A. (1994), *Quality and Reliability of Technical Systems,'* Second Edition, Springer-Verlag

Birolini, A. (1997), *Quality and Reliability of Technical Systems,'* Second Edition, Springer-Verlag.

Blanchard, B S., and Fabrycky, W J. (1991), *Systems Engineering and Analysis,* Prentice Hall, New York.

Blanchard, B.S., Dinesh, V., and Peterson, E. L. (1995), *Maintainability: A key to Effective Serviceability and Maintenance Management*, John Wiley & Sons, New York.

Blanchard, B.S., and Fabrycky, W. J. (1999), *Systems Engineering and Analysis* (Third Edition),' Prentice Hall, New York.

Blanchard, B.S., and Fabrycky W. J. (1991), *Life-Cycle Cost and Economic Analysis*, Prentice Hall, New York.

Bowen, D. B., and Headley, D. E. (2002), Airline Quality Rating 2002, Available at URL http://webs.wichita.edu/?u=aqr.

Box, G. E. P. (1966), Use and Abuse of Six Sigma, *Technometrics,* 8 (4), 625-629.

Brain, P. (1995), *Health and Usage Monitoring Systems into Helicopter Support process*, Project Report, University of Exeter, UK

Breyfogle, F W. (2003), *Implementing Six Sigma: Smarter Solutions Using Statistical Methods*, John Wiley and Sons, New Jersey.

British Standard BS 4778, (1991), Section 3.2; *Glossary of International Terms*, British Standard Institutes, London.

BS 5760 Reliability of systems, equipment and components, (1997), Part 23, *Guild to life cycle costing*. British Standards Institution, London.

Bull, J. W. (1993), *Life Cycle Costing for Construction*, Chapman & Hall, London.

Cini, P.F., and Griffith, P. (1999), Designing for MFOP: Towards the Autonomous Aircraft, *Journal of Quality in Maintenance Engineering*, 5(4), 1999.

Chitra T., and Srinivasan N. K. (1996), System Reliability & Performance Analysis for a sonar, *Proceedings of National Systems Conference NSC-96, VSSC Trivandrum, India*

Chitra, T. (1997), Hazard analysis and its implication for system analysis, *Proceedings of National Symposium on System Analyses for Defence*, CASSA, Bangalore, India.

Chitra, T. (1997), Quality Improvement through Reliability Analysis & Assessment, *National Conference on Quality & Reliability*, NCQR-97, SJCE, Mysore

Chitra, T. (1998), Flight Criticality analysis of AEW Radar using reliability analysis, *Proceedings of National Systems Conference, NSC-97, RCI Hyderabad, India*

Chitra, T. (1998), Reliability prediction and safety analysis for radiation measuring instruments – A Case Study, *National seminar on Reliability Analysis and Engineering,* NSRAE-98, CASSA, Bangalore

Chitra, T. (1998), Availability of preamplifier system in a sonar, *Proceedings of International Seminar on System Operational Effectiveness*, MIRCE, Exeter, UK.

Chitra, T. (2001), Steady State Availability of preamplifier system, *IETE Journal,* 18, (1), 11-15.

Chitra, T. (2002), Constant Failure Rate – Does it Exist ? for a system with Series-Parallel combinations, *Journal of DRDO Reliability and Quality Quest,* 4. (1), 40-43.

Chitra, T. (2002), Data collection for Reliability Analysis, *Journal of Reliability and Quality Quest*, 4, (2), 35-37

Chitra, T. (2002), Risk Analysis of Software through FMEA, *Proceedings of National Seminar on Maintainability of Avionics Systems at Naval Aircraft Yard*, Kochi . India

Chitra, T. (2003), Life Based Maintenance Policy for Minimum Cost, *Proceedings of RAMS-2003*, International Symposium at TAMPA, FL, USA

Chorley, E. (1998), *Field Data – A Life Cycle Management Tool for Sustainable in-Service Support*, M.Sc. Dissertation, University of Exeter, UK

Crocker, J., Dinesh Kumar, U., and Knezevic, J., (1999), Age-Related Maintenance versus Reliability Centered Maintenance: A Case Study on Aero-Engines, *Reliability Engineering and System Safety,* 67, 113-118.

Cole, G.K. (2001), Using Proportional Hazards models, the Cumulative Hazard Function, and Artificial Intelligence techniques to analyse batch problems in engineering failure distributions, *Dissertation submitted for the degree of MSc in Systems Operational Effectiveness*, Centre for M.I.R.C.E., University of Exeter

Cox, D R. (1962), *Renewal Theory,* Methuen & Co Ltd, London.

Creveling, C. M., Slutsky, J. L., and Antis, D. (2003), *Design for Six Sigma,* Pearson Eduction, New Delhi.

Dai, S and Wang M. (1992) *Reliability Analysis in Engineering Applications*, Van Nostrand Reinhold, New York.

Dinesh Kumar, U., and Crocker, J. (2003), 'Maintainability and Maintenance – A Case Study on Mission Critical Aircraft and Engine Components in *Case Studies in Reliability and Maintenance*', (Editors Blischke, W and Murthy D N P), John Wiley and Sons, Hoboken, New Jersey.

Dinesh Kumar, U. (2003), Six Sigma – Implementing Management Strategy Using Mathematical and Statistical tools, *proceedings of the second national conference on Mathematical and Computational Models*, December 11-12, 2003, Coimbatore, India

Dinesh Kumar, U., and Hinds, P. A. (2002), High Level Reliability Impact Analysis – A Case Study on Armoured Vehicle,' *R & D Quality Quest*, 4.(1).

Dinesh Kumar, U. (2001), Setting Reliability Goals for Future Air Systems, *IETE Journal of Research – Special Issue on Quality Management,* 18 (1),. 5-9.

Dinesh Kumar, U., Crocker, J., Knezevic, J., and El-Haram, M. (2000), *Reliability, Maintenance and Logistic Support – A Life Cycle Approach,* Kluwer Academic Publishers, Massachusetts, USA

Dinesh Kumar, U., Crocker, J., and Knezevic, J. (1999), Maintenance free Operating Period – An Alternative Measure to MTBF and Failure for Specifying Reliability, *Reliability Engineering and System Safety*, 64: 127-131

Dinesh Kumar, U. (1999), New Trends in Aircraft Reliability and Maintenance Measures, *Journal of Quality in Maintenance Engineering*,' 5: 287-295.

Dinesh Kumar, U., Crocker, J., and Knezevic, J. (1999), Evolutionary Maintenance for Aircraft Engines, *Proceedings of the Annual Reliability and Maintainability Symposium*, 62-68.

Dinesh Kumar, U., and Birolini, A. (1999), Approximate Expressions for Availability of Repairable Series-Parallel Structures, *International Journal of Reliability, Quality and Safety Engineering*, 6 (4), 319-333.

Dinesh Kumar, U. (1998), Reliability Analysis of Fault Tolerant Recovery Blocks, *OPSEARCH*, 35: 281-294.

Dinesh Kumar, U., and Gopalan, M N. (1997), Analysis of Consecutive k-out-of-n:F Systems with Single Repair Facility, *Microelectronics and Reliability*, 37: 587-590.

Drenick, D F. (1960), 'The failure law of complex equipment, *Journal of the Society for Industrial Applied Mathematics*, 8: 680-690.

Dubi, A. (1999), *Monte Carlo Applications in Systems Engineering*, John Wiley & Sons, Chichester, UK

Ebeling, C. (1997), *An Introduction to Reliability and Maintainability Engineering*,' McGraw Hill International Editions, New York.

El-Haram, M., and Knezevic, J. (1995), *Indicator and Predictor as Two Distinct Condition Monitoring Parameters*, *Proc. of the 5th International Logistics symposium*, pp 179-186, Exeter, UK.

Flanagan, R., and Norman G. (1983), *Life-Cycle Costing for Construction*, Surveyors Publications Ltd.

Fuqua, N B (1987), *Reliability Engineering for Electronic Design*, Marcel Dekker, New York

George, M L (2002), *Lean Six Sigma: Combining Six Sigma Quality with Lean Speed*, Tata McGraw Hill, New Delhi.

Gnedenko, B.V., Belyayev, K., and Solovyev. (1969), *Mathematical Methods of Reliability Theory,'* Academic Press, New York.

Griffith, G. (1998), A Study into the Effects of Reliability and Maintainability on Future Combat Aircraft Logistic Support Costs, *M.Sc. Dissertation,* University of Exeter, UK

Grosh, D L (1989), *A Primer for Reliability Theory*, John Wiley and Sons, New York

Grosh, D L (1982), A Parallel System of CFR Units is IFR, *IEEE Transactions on Reliability,* 31(4), 403-407.

Hahn, G J., Hill, W. J., Hoerl, R W., and Zinkgraf, S A. (1999), The Impact of Six Sigma Improvement – A Glimpse into the Future of Statistics, *The American Statistician*, 53 (3), 208 – 215.

Harry, M. (1998), Six Sigma: A Breakthrough Strategy for Profitability, *Quality Progress,* 60-64.

Henley, E J., Kumamoto, H (1993), *Reliability Engineering and Risk Assessment,* Prentice Hall, New Jersey.

Hess, J A. (1988), Measuring Software for its Reuse Potential, *Proceedings of the Reliability and Maintainability Symposium,* 202-207

Hinds, P. A., and Dinesh Kumar, U. (2001), A Methodology for Assessing the Cost of Failure, *IETE Journal of Research – Special Issue on Quality Management,* 18 (1), 71-76.

Hobbs, G (2001) *'Accelerated Reliability Engineering : HALT and HASS',* Wiley, USA

Hockley, C.J., and Appleton, D.P. (1997), Setting the Requirements for the Royal Air Force's Next Generation Aircraft, *Annual Reliability and Maintainability Symposium,* 44-49.

Kapur, K.C., and Lamberson, (1977), *'Reliability Engineering in Engineering Design,* John Wiley & Sons, New York.

Knezevic, J., (1987a), Condition Parameter Based Approach to Calculation of Reliability Characteristics, *Reliability Engineering,* 19: 29-39.

Knezevic J., (1987b), Required reliability level as the optimization criterion, *Maintenance Management International*, Elsevier, 6: 249-256.

Knezevic, J. (1993), *Reliability, Maintainability and Supportability – A Probabilistic Approach,* McGraw-Hill Book Company, London

Knotts, R., (1996) *'Analysis of the Impact of Reliability, Maintainability and Supportability on the Business and Economics of Civil Air Transport Aircraft Maintenance and Support, M.Phil. Thesis,* University of Exeter, UK

Knotts, R. (1999), *Integrated Logistic Support*, M.Sc. Course Notes, University of Exeter, UK

Knowles, D.I. (1995), Should we Move Away from the Acceptable Failure Rate, *Communications in Reliability, Maintainability and Supportability,* 2: 23-28

Koss, E. (1988), Software Reliability Metrics for Military Systems, *Proceedings of the Reliability and Maintainability Symposium,* 190-194.

Lam., M. (1995), An Introduction to Airline Maintenance, *Handbook of Airline Economics, Aviation Week Group*, 397-406.

Leitch, R D (1986), *Reliability Analysis for Engineers – An Introduction*, Oxford University Press.

Lewis, C. (1871), *Through the Looking Glass and What Alice Found There,* Macmillan, London

Lloyd, D. K., and Lipow, M. (1962), *Reliability: Management, Methods and Mathematics,* Prentice Hall, New York.

Mann, Jr, L., Saxena, A., and Knapp G.M. (1995), Statistical-based or Condition-based Preventive Maintenance, *Quality in Maintenance Engineering*, 1: 46-59.

Matt, R. (1993), *The Red Queen: Sex and the Evolution of Human Nature,* Penguin.

MIL-STD 721C. (1966), *Military Standard, Definitions of Effectiveness Terms for Reliability, Maintainability, Human Factors and Safety*, Department of Defense, Washington, DC.

MIL-STD-1629A (1984), *Procedures for Performing a Failure Mode, Effects and Criticality Analysis*, Department of Defense, Washington D.C.

MIL-STD-882C (1993), *System Safety Program Requirement*', Department of Defense, Washington D.C

Misra, K.B. (1991), *Reliability Analysis and Prediction – A Methodology Oriented Treatement,* Elsevier, Amsterdam.

Mobley, R.K. (1990), *An Introduction to Predictive Maintenance*, Van Nostrand Reinhold, New York.

Morgan, P. (1999), *No Fault Found – The Human Factors and the Way Forward, Project Report*, University of Exeter, UK.

Moubray, J. (1997), *Reliability-centred Maintenance - RCM II*, Butterworth and Heinemann, Oxford.

Nevell, D. (2000), Using Aggregated Cumulative Hazard Plots to Visualize Failure Data, *Quality and Reliability Engineering International,* 16(2), 209-219.

O'Connor, P.D.T. (2004), *Practical Reliability Engineering*, 4th edition, John Wiley & Sons,UK

Palm, C. (1938), Analysis of Erlang Formulae for Busy Signal Arrangements, *Ericsson Techniques,* 4: 39-58.

Pearcy, D. (1999), A Study into Operational Life Cycle Costs by Investigating the Interaction Between 'Out-of-Service' Time and the Maintenance Concept, *M.Sc Dissertation,* University of Exeter, UK

Rajeha, D G. (1991), *Assurance Technologies: Principles and Practices*, McGraw Hill, New York.

Rosenberg, L. H., Hammer, T.E.D., and Shaw, J. (1998), Software Metrics and Reliability, *Proceedings of the 9th International Symposium on Software Reliability Engineering,* Germany.

Saranga, H. (2000), *Relevant Condition Parameter Based Approach To Reliability And Maintenance*, PhD Thesis, Centre for M.I.R.C.E., University of Exeter, UK

Saranga, H and Knezivic, J. (2000) Reliability Analysis Using Multiple Relevant-Condition Parameters", Journal of Quality in Maintenance Engineering, MCB University Press, 6 (3), 165-176.

Saranga, H and Knezevic, J. (2001), Reliability Prediction for Condition Based Maintained Systems, *Reliability Engineering & System Safety*, 71, 219-224.

Saranga, H.(2002), Relevant Condition Parameter Strategy as an effective Condition based Maintenance, *Journal of Quality in Maintenance Engineering*, 8, 92-105.

Saranga, H and Dinesh Kumar U (2006), *Optimization of Aircraft Maintenance and Support Infrastructure: Level of Repair Analysis*, Forthcoming *Annals of Operations Research.*

Schmidt, S. R., and Launsby, R. G. (1997), *Understanding Industrial Designed Experiments,* Air Academy Press, Colarado, CO.

Srinivasan, N. K., and Chitra T. (1999), Reliability Engineering Programmes Some Perspectives, *Journal of Reliability & Quality Quest*, 1.(1), 17-19.

Srinivasan, N. K., and Chitra T. (1999), Reliability Analysis for an Indigenously developed system – A Case study, *Journal of Reliability & Quality Quest*, 1 (2), 6-27.

Stamatis, D. H. (2001), *Six Sigma and Beyond – Foundations of Excellent Performance,* St. Lucie Press.

Tijms, H. (1995), *Stochastic Models – An Algorithmic Approach,* John Wiley, New York.

Thawani, S. (2004), Six Sigma – Strategy for Organizational Excellence, *Total Quality Management*, 15, 655-664.

Turner, M. (1999), Unmanned Compact Aircraft, *project Report,* University of Exeter, UK

US MIL-HDBK-217, *Reliability Prediction for Electronic Systems,* USAF, Rome

Valdes-Flores, C., and Feldman R. (1989), A Survey of Preventive Maintenance Models for Stochastically Deteriorating Single-Unit Systems, *Naval Research Logistics Quarterly* 36, 419-446.

Verma D., and Knezevic. J. (1995), Conceptual System Design Evaluation: Handling the Uncertainty,' *Proceedings of the International MIRCE Symposium*, Exeter, 118-130.

Verma, D. (1993), A Causal Emphasis During The Failure Mode, Effects and Criticality Analysis,' *Proceedings of the International Logistics Symposium,* 120-126.

Williams, J.H., Davies, A., and Drake, P.R. (1994), *Condition-based Maintenance and Machine Diagnostics*, Chapman & Hall, London.

Xie, M. (1989), On the Solution of Renewal-Type Integral Equations, *Communications in Statistics, B.,* 18: 281-293.

Xie, M. (1991), *Software Reliability Modelling,* World Scientific Publishing Company, Singapore.

Index

A

B

C

I

J

K

L

M

N

O

P

Q

R

S

T

U

V

W

Y